Transflective Liquid Crystal Displays

Wiley-SID Series in Display Technology

Series Editor:
Anthony C. Lowe

Consultant Editor:
Michael A. Kriss

Transflective Liquid Crystal Displays

Zhibing Ge & Shin-Tson Wu

College of Optics and Photonics
University of Central Florida, USA

A John Wiley and Sons, Ltd., Publication

This edition first published 2010
© 2010, John Wiley & Sons, Ltd

Registered office
John Wiley & Sons Ltd, The Atrium, Southern Gate, Chichester, West Sussex, PO19 8SQ,
United Kingdom

For details of our global editorial offices, for customer services and for information about
how to apply for permission to reuse the copyright material in this book please see our
website at www.wiley.com.

Library of Congress Cataloguing-in-Publication Data

Ge, Zhibing.
  Transflective liquid crystal displays / Zhibing Ge, Shin-Tson Wu.
     p. cm.
  Includes bibliographical references and index.
  ISBN 978-0-470-74373-7 (cloth)
  1. Liquid crystal displays. 2. Reflective materials. I. Wu, Shin-Tson. II. Title.
  TK7872.L56G425 2010
  621.3815′422–dc22

                                                        2009049255

A catalogue record for this book is available from the British Library.

ISBN: 978-0-470-74373-7 (Hbk)

Set in 10/12.5pt, Palatino by Thomson Digital, Noida, India
Printed in Great Britain by TJ International Ltd, Padstow, Cornwall

Contents

Series Editor's Foreword

The requirements for flat panel displays are ever evolving towards lower power, more saturated colours, higher reflectivity, greater luminance, faster response, wider viewing angle *etc*. Yet there is one area of display technology, the subject of the present volume in the Wiley-SID Series in Display Technology, where any such improvements become the more difficult to achieve because of the need for displays to operate under any ambient illuminant condition from darkness to bright sunlight. Clearly, the image on a reflective display will be invisible in darkness; that on a backlit or self-luminous display will wash out in full sunlight.

Thus we enter the world of the transflective display – one which is partially backlit and partially reflective. This architecture offers many challenges to the display engineer; a transflective display can never achieve the luminance of a backlit display nor the reflectivity of a reflective display, so compromise will always be involved. Such compromise can be minimised by innovative design of liquid crystal electro-optic effects so that the threshold voltage, the gamma curve, the chromatic properties, the off-axis viewing characteristics and the cell thickness requirements of the transmissive part of the display match as closely as possible those of the reflective part.

In their preface, the authors have described in detail the contents of the individual chapters of this book, so I shall not dwell on that in more detail here. However, I would emphasise that transflective displays require some of the most elegant and exacting applications of liquid crystal science. That fact is demonstrated abundantly in this book, which covers the basic concepts, the device physics and provides detail of every known liquid crystal effect applicable to transflective displays at a level suitable for both postgraduate students and practising engineers. I thank the authors for their efforts in writing this timely addition to the series.

Anthony Lowe
Series Editor
Braishfield, UK, 2010

About the Authors

Zhibing Ge

Zhibing Ge received his BS degree in electrical engineering from Zhejiang University, Hangzhou, China in 2002, and MS and PhD degrees in electrical engineering from the University of Central Florida (UCF), Orlando in 2004 and 2007, respectively. Currently, Dr Ge is a senior display engineer in the Flat Panel Displays department at Apple Inc., California. Between January 2008 and August 2009 he was with the College of Optics and Photonics, University of Central Florida, as a research scientist. His research interests include novel liquid crystal displays and laser beam steering technologies. He has published a chapter in a book, over 30 journal papers, and 12 issued or pending patents in related areas. Dr Ge is the recipient of the 2008 Otto Lehmann award. Since May 2008, he has been serving as an associate editor for the *Journal of the Society for Information Display* (*JSID*) in the LCD division.

Shin-Tson Wu

Shin-Tson Wu is a PREP professor at the College of Optics and Photonics, University of Central Florida (UCF). Prior to joining UCF in 2001, he worked at Hughes Research Laboratories (Malibu, California) for 18 years. He received a PhD in physics (quantum electronics) from the University of Southern California (Los Angeles, California, USA) and a BS in physics from the National Taiwan University (Taipei, Taiwan).

Professor Wu was the recipient of the 2008 SID Jan Rajchman prize and the 2008 SPIE G. G. Stokes award. He is a Fellow of the IEEE, SID, OSA, and SPIE. He has co-authored five books: *Introduction to Flat Panel Displays*

(Wiley, 2008), *Fundamentals of Liquid Crystal Devices* (Wiley, 2006), *Introduction to Microdisplays* (Wiley, 2006), *Reflective Liquid Crystal Displays* (Wiley, 2001), and *Optics and Nonlinear Optics of Liquid Crystals* (World Scientific, 1993), six book chapters, over 350 journal papers, and 66 issued patents. Professor Wu was the founding Editor-in-Chief of the IEEE/OSA *Journal of Display Technology*.

Preface

The explosive growth in personal mobile electronics has not only driven continuous improvement in image quality and power efficiency of existing displays, but also has nourished rapid development of novel ones that could have multiple functions. A stylish display has become a key factor for consumers when choosing an electronic device. Among various display technologies such as organic light-emitting displays (OLEDs), super-twisted nematic (STN), liquid crystal displays (LCDs), and electronic ink (e-ink), active matrix LCDs dominate the mobile display market owing to their capacity to provide high resolution and full color with high contrast and a wide viewing angle, and good legibility under different ambient conditions. A characteristic of LCDs in terms of maintaining vitality in the market is our ability to make advances in most of their components, from the backplane backlight unit to the front polarizer and anti-reflection film.

With the rising demand for outdoor image readability in portable displays, two parallel research and development directions in novel LCDs are: (i) developing high brightness transmissive LCDs that adaptively adjust the backlight output according to different ambient light conditions, and (ii) developing transflective LCDs that incorporate both transmissive and reflective functions into one display. The first approach maintains the superior image quality and simple device architecture of the transmissive mode, but its power consumption will be too high even for a short time use to enable it to compete with reflected sunlight, resulting in a short battery life. Transflective LCDs, on the other hand, utilize sunlight as an external light source for outdoor applications, thus their power consumption is relatively low. In addition, their outdoor ambient contrast ratio can be almost independent of the lighting conditions to maintain good image legibility. But the image quality of the transmissive sub-pixel might be compromised by, for example, a reduced transmissive aperture and lower contrast ratio. Both approaches have their own merits and demerits. However, the performance of a transflective LCD can still be improved, to ultimately have comparable image quality to a transmissive LCD in a dim environment as well as high reflectivity, high contrast, and good color reproduction in the reflective sub-pixel for outdoor use. This requires engineers and scientists to first have an in-depth understanding of the device architecture, operating principles, and device performance. This is the major purpose of this book.

Chapter 1 gives a general review of LCD systems and key display elements covering the backlight unit, various LC modes, reflectors, linear polarizers, and compensation films. This chapter serves as a basis for readers to become familiar with transmissive and transflective LCDs.

Chapter 2 describes the device physics and modeling methods associated with the design and characterization of, but not limited to, typical transflective LCDs. In the modeling methods, we delve into LC molecular reorientation under the influence of an electric field and the corresponding electro-optical properties. Several exemplary transflective LCD device configurations are used to illustrate the design principles and underlying physics.

Chapters 3 and 4 further describe the wide-view technologies which are crucial to both transmissive LCD TVs and transflective LCDs. In Chapter 3, the detailed optical principles associated with phase compensation are illustrated using a Poincaré sphere. The analysis of linear polarizer and circular polarizer compensation provides readers with a clear pattern of the status and challenge in achieving wide-view transflective LCDs, especially in multi-domain vertical alignment (MVA) mode-based devices. In Chapter 4, detailed device configuration and design considerations of MVA, in-plane switching (IPS), and fringe field switching (FFS) transflective LCDs are discussed. Up-to-the-minute progress in these technologies is introduced, such as polymer-sustained surface alignment technology for MVA LCDs and in-cell-retarder-free FFS transflective LCDs. These two technologies possess great potential for next generation mobile displays.

Chapter 5 discusses the fast-response technology intended for color sequential displays and video-rate transflective LCDs. The fundamental physics and electro-optics including flow, response time, and compensation of optically compensated bend (OCB) modes using a pi-cell to achieve a fast response time are analyzed in detail.

Chapter 6 is designed to give readers some future technological perspectives. The comparison of a backlit transmissive display with a transflective

LCD is investigated in the real environment from which readers can see the irreplaceable role of transflective technologies in mobile displays. Emerging functionalities such as touch panel technology, including both external and in-cell touch designs, are also briefly introduced.

By the nature of their inherent advantages and great improvement potential, transflective LCDs will continue to attract significant research and application attention. This book aims to provide readers with a comprehensive introduction to the device configuration, underlying device physics, technical design considerations, and technological perspectives of transflective LCDs for next generation, sunlight-readable mobile display applications.

We would like to thank those who provided technical and financial support during the preparation of this book. In particular, we are deeply grateful to Dr Xinyu Zhu (Dow Chemical), Professor Thomas X. Wu (University of Central Florida), Dr Chung-Kuang Wei and Dr Wang-Yang Li (Chi-Mei Optoelectronics, Taiwan), Professor Seung Hee Lee (National Chonbuk University, Korea), and Dr Hyang-Yul Kim (Samsung, Korea), and our LCD group members. We are also grateful to Series Editor Anthony Lowe for his valuable suggestions and proofreading of the book. Special thanks go to our family members. With their understanding and spiritual support, we were able to finish this book on time.

Zhibing Ge, California, USA
Shin-Tson Wu, Florida, USA

1

Device Concept of Transflective Liquid Crystal Displays

1.1 Overview

The ability to electrically tune optical birefringence makes liquid crystal (LC) a useful material for electro-optical applications. Because of its compact size and light weight, liquid crystal displays (LCDs) have been used extensively in electronic devices and appliances, from microdisplays and small handheld mobile phones to medium-sized notebook and desktop computers and large-panel LCD TVs. After decades of development, LCD technology is capable of giving excellent image performance at a relatively low cost. Compared with other display technologies, a unique characteristic of LCDs in terms of maintaining long-term competitiveness is the vigorous technical progress being made in almost every key element, including the LC material, TFT, LC cell structure, color filters, compensation films (including polarizers), backlight source, backlight films, driving electronics and algorithms, etc. In the foreseeable future, LCDs will continue to maintain their dominance in the display market.

Presently, three types of LCD have been developed to suit different applications: (i) transmissive, (ii) reflective, and (iii) transflective LCDs. In a transmissive LCD, a backlight is usually embedded as the light source to illuminate the LCD panel. A transmissive LCD typically exhibits high

Transflective Liquid Crystal Displays   Zhibing Ge and Shin-Tson Wu
© 2010 John Wiley & Sons, Ltd

brightness (300–$500\,cd/m^2$), high contrast ratio ($>1000:1$), and good color saturation. With its high image quality, transmissive LCD is the most adopted display technology and its performance is being improved steadily. For example, with film compensation and a multi-domain structure, a contrast ratio over $50:1$ can now be easily obtained omnidirectionally in a wide-view LCD TV. Utilizing new backlight sources, like RGB light emitting diodes (LEDs), the color gamut can reach over 110% of NTSC to show pretty rich and vivid colors. With advanced low-viscosity LC materials and fast LC modes, a refresh rate of over $120\,Hz$ (i.e., $240\,Hz$ or even $480\,Hz$) has been achieved. Together with backlight local dimming, the dynamic image qualities, including dynamic contrast ratio and motion picture response time, have also been continuously enhanced which, in turn, leads to a significant energy saving. Therefore, for mobile displays, the high image quality of a transmissive LCD is advantageous, but some of its features mean it is still imperfect for mobile applications. First, transmissive LCDs rely on backlight white LEDs to illuminate the image. The associated power consumption is relatively high, resulting in a short battery life. Second, for some outdoor situations such as under strong sunlight, the surface luminance provided by a backlight cannot compete with the sunlight, leading to a washed out image and poor outdoor readability. Although increasing backlight intensity to elevate surface brightness above the ambient light intensity could improve the readability, the power consumption would also be dramatically increased. A solution for reaching good sunlight readability of a transmissive LCD is to embed an ambient light intensity sensor into a pure transmissive LCD that can dynamically or smartly adjust the brightness of the backlight to meet different ambient conditions. But this approach can only really target short usage times given the power consumption needed. For example, most transmissive mobile displays have a surface luminance at about 200 to $300\,cd/m^2$. To compete with regular outdoor sunlight (direct sunlight $> 100\,000\,lux$), huge backlight power is required to increase surface luminance to over $500\,cd/m^2$, even with anti-reflection coatings on the display front surface.

Another type of display is the reflective LCD, which embeds a metal reflector behind the LC cell, such as a mixed twisted nematic cell [1, 2] or cholesteric cell [2, 3], and uses external light to illuminate the displayed image. A reflective LCD has low power consumption and good sunlight readability. On the other hand, since a reflective LCD relies on external light to display the image, it exhibits poor readability in a low ambient light environment. Besides, the imperfect removal of surface reflection degrades its contrast ratio and color performance. Its low reflectivity (in most cases using circular polarizers), low contrast ratio (a typical value is $\sim5:1$, while a diffusive white paper has a contrast ratio of ~15 : 1) and low color saturation give viewers a different image experience from what would be seen on a transmissive LCD. Therefore, reflective LCDs are usually employed for low-end applications, such as devices that only require outdoor daytime viewing. Nevertheless, their low power consumption and sunlight readability are superior to transmissive displays, making reflective LCDs useful in portable devices where battery life is critical.

As a direct consequence, transflective LCDs have been designed to combine both transmissive and reflective functions into one display [4, 5]. In dark or low ambient light conditions, the backlight is turned on and the image is mainly displayed in the transmissive mode, exhibiting excellent image quality with high contrast ratio and good color saturation. In a bright ambient light situation such as under strong sunlight, the reflective mode mainly functions to display images and the backlight may either be turned on to assist the image display or turned off to save power. According to different application requirements, the ratio of transmissive to reflective regions can be varied. For instance, for a portable audio player that requires longer battery life, a larger reflective region can be adopted. But for a video cell phone where a vivid image is very important, the transmissive region can be made larger. Driven by the increasing demands of the mobile electronics market, high brightness, wide viewing angles, vivid color, fast response, outdoor read-ability, and low cost are all now required for small mobile displays. Research into transflective LCDs is being undertaken to meet such requirements.

In this book we will focus on the mainstream TFT-addressed wide-view transflective LCDs for mobile applications. First, the fundamental device elements including polarizers, liquid crystal alignment, relevant compensation films, reflectors, and backlight films that comprise a portable LCD device and their associated principles are briefly introduced. This is followed by the introduction of related modeling methods of LCD directors and optics which serve as the basis for transflective LCD characterization and optimization. More details on developing good light efficiency and wide viewing angle transflective LCDs are then discussed in terms of both TFT LC cell design and compensation films. Advanced topics related to applications like touch panels and fast response time for video mobile displays will also be addressed. Finally, Chapter 6 re-addresses the unique and irreplaceable role of transflective LCDs in the mobile display market and discusses the possible directions in which they could be developed to enhance competitiveness.

Figure 1.1 depicts the device structure of an active matrix-driven trans-missive LCD based on amorphous silicon (a-Si) thin-film-transistor (TFT) technology. An LCD is a non-emissive display, i.e., it does not emit light by

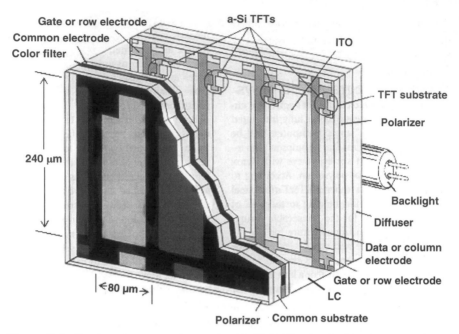

Figure 1.1 Device structure of one pixel (with RGB sub-pixels) of a transmissive TFT LCD

itself; instead, it utilizes a backlight and the LC cell sandwiched between two sheets of stretched dichroic polarizers functions as a light valve. A diffuser close to the backlight is used to homogenize the backlight intensity in order to avoid hot spots. Prism films, such as 3M´s brightness-enhancement-film (BEF) [6] are stacked to confine the incident Lambertian backlight into a central cone to ±40° for enhancing axial display brightness. On the rear substrate (the TFT-array substrate), a TFT-array is formed to provide an independent switch for each pixel. Color filters are fabricated on the front substrate (namely the color-filter-array substrate) and aligned with the rear TFT pixels. In each time frame, signals (short high-voltage pulses) from the gate lines turn on the TFTs in a scanning sequence, and the voltages from the data lines are applied thereafter to drive each individual LC pixel to the targeted gray level. Under such a spatial RGB-sub-pixel configuration, different colors are achieved by combining the separate colors from RGB sub-pixels at assigned gray levels, and the eyes average the overall optical response from them. For large-panel TV applications, the typical sub-pixel size is about $80 \, \mu m \times 240 \, \mu m$. The sub-pixel size is reduced to about $50 \, \mu m$ $150 \, \mu m$ in small cell phone panels. In practice, the effective aperture for light transmission is much smaller than the sub-pixel area. The aperture ratio, defined as the effective region for transmission over the total region of each pixel, is usually less than 80%. Several factors cause a low aperture ratio. For TFTs, the active channel is sensitive to visible light, requiring a light shielding layer to cover that region. To avoid light leakage and color mixing from adjacent sub-pixels, black matrix (BM) is also formed at the boundaries between sub-pixels. The metal or alloy gate and data lines and the storage capacitor, made of opaque material, lead to a further reduction in the aperture ratio. Even for the area that is transparent to light, the light transmission cannot always reach 100% owing to the LC alignment there. In some LC modes such as the multi-domain vertical alignment (MVA) LCD [7–9], non-desired LC reorientations such as disclination lines also lower the effective light output.

For the LC cell, both inner surfaces are coated with a thin ($\sim$80 nm) polyimide layer to provide initial alignment of LC molecules that will adjust the ordered LC reorientation when a voltage is applied. Presently, major LCD technologies like twisted nematic (TN) [10], in-plane switching (IPS) [11, 12], fringe field switching (FFS) [13], and pi-cell or optically-compensated-bend (OCB) cells [14, 15] require a surface rubbing alignment, and technologies based on the VA mode, such as MVA [7–9], and patterned vertical alignment (PVA) [16] can yield initial vertical alignment without rubbing for high contrast. The LC cell gap is usually controlled at about 4.0 µm for a transmissive LCD, and the phase change of the LC layer between a fully bright state and a dark state is about half a wavelength. The electro-optical performance of the display in terms of light efficiency, response time, and viewing angle is related to the LC material, alignment, and cell structure.

For a direct-view reflective LCD, the cross-sectional view is depicted in Figure 1.2. This device utilizes external ambient light to display images, where incident light is reflected by the rear reflector and traverses the LC cell twice. The front linear polarizer and the retardation film (such as a quarter-wave plate) form a crossed-polarizer configuration for incident light. The aperture ratio of a reflective LCD is much higher than that of a transmissive one, since the TFT, and storage capacitor can be buried under the metal reflector. The cell gap of the LC layer is typically about 2 µm and the phase change between the bright state and the dark state is about quarter of a wavelength. Here, the color filter layer is formed on the front substrate in the figure, but alternatively it could be formed on the rear glass substrate. To avoid specular reflection and widen the viewing angle, a bumpy reflector surface is usually required and asymmetrical reflection is preferred, i.e., the incident angle and exit angle of the rays are designed to be different. Thus, strong surface specular reflection, which is considered noise, will not

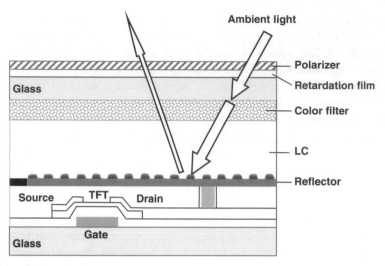

Figure 1.2 Device structure of one sub-pixel of a reflective TFT LCD

overlap with the useful signal coming out of the LC cell. Compared with a transmissive LCD, the contrast ratio, viewing angle, and color saturation of a reflective LCD are inferior, resulting from many factors such as non-negligible surface reflection, complex optical films for circular polarizer configuration, and uncolored openings on the color filters for high reflectivity. Research efforts are being made to improve the image quality of reflective LCDs, e.g., by designing new anti-reflection films and high-reflectivity reflectors.

1.2 Polarizers

1.2.1 Linear Polarizers

As discussed above, optical performance of an LCD relies on each optical element, such as the LC cell, compensation film, and polarizer. For transmissive LCDs, linear polarizers partially determine the contrast ratio and hue balance (the spectral distribution of light output from the polarizers) of the display. For reflective LCDs, circular polarizers, comprising both a linear polarizer and a quarter-wave plate, are usually employed where viewing angle and dark-state spectral light leakage are important. Below, we will briefly introduce different polarizers employed in LCDs and their associated mechanisms. Typically, linear polarizers for LCDs are made from polyvinyl-alcohol (PVA) films with iodine compounds using the wet-dyeing method [17]. After the PVA is stretched, dichroic species, such as I_3^- and I_5^- complexes, are aligned along the stretching direction, thus light polarized along this stretch direction will be strongly absorbed, while light polarized perpendicular to this direction will be transmitted. The degree of polarization and transmittance of PVA-stretched polarizers is highly dependent on the dichroism and the amount of dichroic species. In other words, one can control these parameters to adjust the transmittance and hue balance of the polarizers, which is very important for displays, especially for LCD TVs.

Figure 1.3 shows the optical transmittance from two identical linear polarizers set parallel or perpendicular to each other. For the parallel setup, the output transmittance in the blue region is much weaker than that at longer wavelengths. Here, the absorption of light is proportional to $\exp\left(-\frac{2\pi}{\lambda}dn'\right)$, where λ is the wavelength, d is the thickness of the medium, and n' is the imaginary part of the refractive index of the polarizer along the transmission or absorption axis. In a typical polarizer, the transmission axis n' (of the order of 10^{-5}) decreases as wavelength increases. Blue light (with a smaller λ value) experiences a larger $\exp\left(-\frac{2\pi}{\lambda}dn'\right)$ value and in turn a stronger absorption than red light through two parallel polarizers. As a result, a weaker output of blue light in the bright state causes the so-called *blue decoloration phenomenon*, making the display at full bright state appear a little yellowish. On the other hand, light leakage from these two crossed linear polarizers is inherently well suppressed over most of the visible range, although there is still evident light

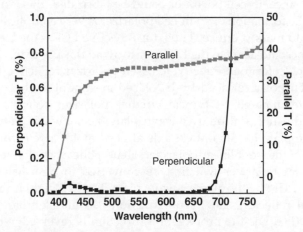

Figure 1.3 Wavelength-dependent transmittance of two linear polarizers at parallel (open) and perpendicular (closed) positions

leakage in the blue and red regions, as shown in the figure, causing decoloration of both bluish and reddish colors in the dark state. To eliminate or reduce decoloration for a better color balance, material engineering is applied. Each dichroic species has unique absorption peaks at different wavelengths and we can control the I_3^- and I_5^- complexes to increase the absorption in the blue region to suppress bluish decoloration in the dark state. Sometimes compensation films, including optimized TAC films, are also helpful in reducing the color shift of the dark state [17, 18]. For the bright state color balance, in addition to optimizing the material properties, we can adjust the transmittance of each color via backlight or LC cell engineering (controlling the output transmittance of each sub-pixel) to compensate for the weak blue color, achieving a balanced white point.

1.2.2 Circular Polarizers

The polarizer configuration for reflective or transflective displays, however, is more complicated than that of a transmissive display. Previously, for reflective LCDs used in low-end products like wrist watches, two crossed linear polarizers were employed to sandwich the LC layer and a metal reflector or transflector was formed behind the rear polarizer [5, 19]. Since the total thickness of the rear polarizer (including protective layers) and the LC substrate is much larger than the LC cell gap and the pixel size, parallax occurs, which creates a shadow image when the display is viewed at oblique angles, thus degrading the image quality of the display [5]. Of course, if a high-performance, in-cell polarizer could be fabricated, this issue would be solved. But the formation of an in-cell polarizer is very complicated and they are not easy to mass produce. Thus, for reflective LCDs, the metal reflector should be placed adjacent to the LC layer to avoid this parallax. Under such a configuration, to form an effective crossed polarizer reflective mode structure, a circular polarizer needs to be placed in front of the LC cell. The rear polarizer is eliminated. A typical circular polarizer consists of a linear polarizer and a uniaxial quarter-wave plate, as shown in Figure 1.4. The quarter-wave plate has its optical axis aligned at 45° away from the transmission axis of the front linear polarizer. Hence, linearly polarized light from the front linear polarizer will first be converted to right-handed circular polarization (RCP) by the quarter-wave plate. The RCP light is then reflected by the metal reflector, and its polarization becomes left-handed circular polarization (LCP) as the propagation direction is reversed. Note the light maintains its absolute rotation in the lab x-y-z coordinate sense, as shown in Figure 1.4. The LCP light is then converted by the quarter-wave plate into

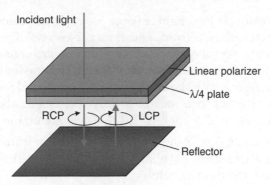

Figure 1.4 Configuration of a circular polarizer using a linear polarizer and a quarter-wave plate

linearly polarized light perpendicular to the front linear polarizer transmission axis, and is blocked. Using circular polarizers in a transmissive LCD based on the MVA mode can also greatly increase the light efficiency owing to the removal of azimuthal angle dependence of LC directors [20, 21].

The configuration shown in Figure 1.4 is simple but has the following issues: (i) high spectral light leakage, and (ii) high off-axis light leakage. The quarter-wave plate has quarter-wave retardation only at a single wavelength (e.g. 550 nm), and it is incapable of covering the whole visible spectrum. Figure 1.5 illustrates wavelength-dependent light leakage at normal and at

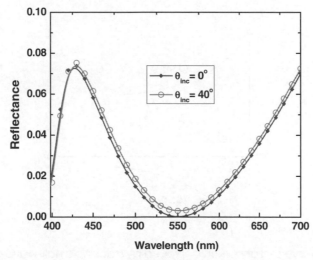

Figure 1.5 Spectral light leakage from a monochromatic circular polarizer

40° incidence. Relatively high light leakage occurs at wavelengths other than the central wavelength of 550 nm, and it increases at oblique incidence. In addition to the imperfect absorption of the linear polarizer used (as shown in Figure 1.3), another important reason for the high spectral light leakage of the configuration in Figure 1.4 is the wavelength dispersion of the quarter-wave plate. Here the birefringence (Δn) of a polymeric retardation film can be roughly estimated as $\Delta n \propto A + \frac{B}{\lambda^2 - \lambda_0^2}$ [17, 22], where A and B are constants, and λ_0 is the absorption edge wavelength and is usually in the ultraviolet range. As the wavelength λ increases, Δn will decrease and $d\Delta n/\lambda$ in turn decreases farther away from the desired value.

From the above analysis, the first method to suppress spectral light leakage is to design new quarter-wave plates with a reverse birefringence (Δn) dispersion by stacking plates with different Δn dispersions, i.e., $d\Delta n/\lambda$ remains roughly a constant throughout the visible range [17, 18]. For example, a quarter-wave plate of reverse dispersion can be manufactured based on subtraction by stacked half-wave and quarter-wave plates with different wavelength dispersions, while their optical axes are set perpendicular to each other. A rough illustration of this method is shown in Figure 1.6. This approach is simple, but the overall dispersion obtained from two real polymeric materials might still be far from an ideal reverse dispersion curve, where $d\Delta n/\lambda$ would remain constant throughout the visible range.

A second method is to stack several retardation plates with different optical axis alignment to self-compensate the wavelength dispersion. Unlike the first

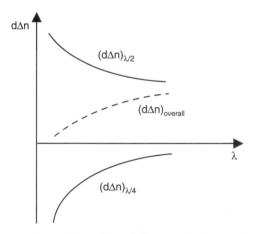

Figure 1.6 Reversed dispersion achieved by stacking a half-wave plate and a quarter-wave plate with different dispersion relations

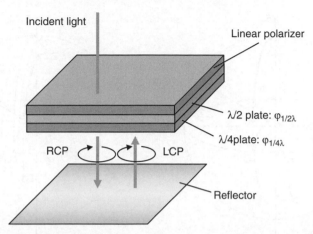

Figure 1.7 Broadband circular polarizer achieved by stacking a linear polarizer, a half-wave plate, and a quarter-wave plate

method, here the retardation plates can have either the same or different dispersion. Using this second approach, it is easier to achieve good dispersion compensation and this method is now widely adopted in practical circular polarizer designs. As shown in Figure 1.7, the broadband circular polarizer consists of a linear polarizer (transmission axis at 0°), a monochromatic half-wave plate (optical axis at $\varphi_{1/2\lambda}$), and a monochromatic quarter-wave plate (optical axis at $\varphi_{1/4\lambda}$) [21]. Here the relation of these angles can be demonstrated from the polarization change trace on the Poincaré sphere [23, 24]. Figure 1.8(a) illustrates the polarization trace of the incident light at a designated wavelength on the Poincaré sphere. Assuming the transmission axis of the linear polarizer is at point **T** on the Poincaré sphere, the optical axis of the first half-wave plate is at point **H**, with $\angle TOH = 2\varphi_{1/2\lambda}$, and the optical axis of the quarter-wave plate is at point **Q**, with $\angle TOQ = 2\varphi_{1/4\lambda}$ (here the relative angle of the optical axis represented on the Poincaré sphere is twice the absolute value in x-y-z coordinates). The linearly polarized light from the front linear polarizer is first rotated from point **T** along **OH** by half a circle when traversing the first half-wave plate; thus, its polarization goes to point **C** at $\angle TOC = 4\varphi_{1/2\lambda}$. The next quarter-wave plate converts the linear polarization from point **C** to RCP on the north pole, i.e., the $\angle QOC$ should be 90°. Hence, $2\varphi_{1/4\lambda} - 4\varphi_{1/2\lambda} = 90°$ is required for this configuration to function as a circular polarizer. Moreover, to obtain broad bandwidth operation, the value of $\varphi_{1/2\lambda}$ is also very important. Figure 1.8(b) roughly depicts the polarization change of different wavelengths (R: 650 nm, G: 550 nm, and B: 450 nm, where each retardation film thickness is designed only for 550 nm). The retardation value $d\Delta n / \lambda$ is inadequate for red

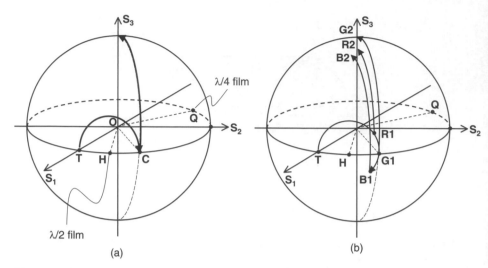

Figure 1.8 Polarization traces on the Poincaré sphere from the broadband circular polarizer for (a) green light (550 nm) and (b) R, G, and B light at normal incidence

light, but too high for blue light. As a result, after the first half-wave plate, the polarizations for red, green, and blue light locate at points **R1**, **G1**, and **B1**, respectively. The location of red light at **R1** is closest to the north pole, thus the subsequent inadequate phase from the quarter-wave plate can still convert light close to circular polarization at the north pole. Similar compensation also occurs for blue light. From this analysis, in addition to the above alignment relation, $\varphi_{1/2\lambda}$ is also very important for such self-compensation of the wavelength. An optimized angle for $\varphi_{1/2\lambda}$ is about 15°, thus $\varphi_{1/4\lambda}$ is about 75° [21]. To achieve broadband operation, the polarization traces from the half-wave plate and the quarter-wave plate should stay in the same upper or lower hemisphere.

The spectral light leakage from the configuration in Figure 1.7 can be greatly suppressed at normal incidence. However, this design still exhibits off-axis light leakage, and with more retardation films, its off-axis performance could be even worse than the above-mentioned monochromatic circular polarizer at the center wavelength, as shown in Figure 1.9. For mobile reflective or transflective LCDs using circular polarizers, a wide viewing angle is also critically important. Especially for transflective LCDs, the transmissive sub-pixel will be sandwiched between circular polarizers, and high off-axis light leakage is not acceptable. Additional compensation schemes are needed to achieve both a wide viewing angle and broadband operation in circular polarizers, which will be discussed in detail in Chapter 3.

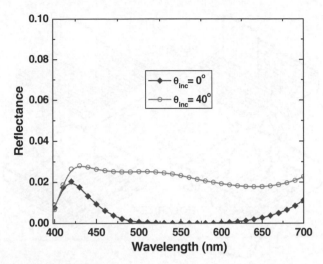

Figure 1.9 Simulated light leakage from the broad-band circular polarizer at normal and oblique angles

1.3 LC Alignment

To be used in display devices, LC mixtures need to be confined and aligned with specific pre-orientation between two substrates made from, for example, glass or plastic. Three commonly used alignment methods are mechanically rubbed polyimide (PI), ion beam etched PI, and evaporated SiO_x [25]. With its simple fabrication process, rubbed PI is the most popular alignment method. Inorganic SiO_x based alignment is mainly adopted in projection displays owing to its robustness and ability to withstand high-intensity illumination.

The initial surface alignment of LCs determines the electro-optical properties of the LC device. For initial molecular distribution, various LC cell alignment technologies, including homogeneous alignment, twisted nematic alignment, vertical alignment, pi-cell alignment, and hybrid alignment, have been developed [4]. Below, we will briefly address the main characteristics and applications of several mainstream LC alignment technologies.

1.3.1 Twisted Nematic (TN) Mode

Twisted nematic alignment [10] is the most commonly used liquid crystal technology for small and medium-sized LCD panels. The basic configuration of a TN cell with LC director profile at the bright (voltage-off) and dark

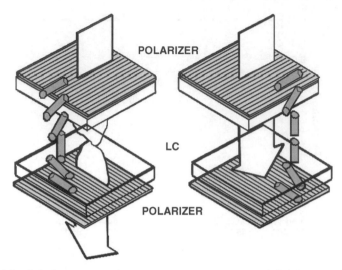

Figure 1.10 Configuration of a TN cell in the voltage-off (left) and voltage-on (right) states

(voltage-on) state is shown in Figure 1.10. The front and rear LC surfaces are rubbed at an angle difference of 90°, thus the LC optical axis gradually twists by 90° from rear to front when no voltage is applied. When a relatively high voltage is applied, the vertical electric field between the rear and front substrates makes the LC directors tilt vertically in the bulk region, as shown in Figure 1.11, where ϕ and θ stand for the azimuthal and polar angles, respectively. When a relatively high voltage is applied (such as $V = 5V_{\text{th}}$, where V_{th} is the threshold voltage), the whole LC cell seems to consist of three parts: a thick vertical-alignment-like cell in the central bulk region with most directors tilted vertically, and two crossed thin hybrid-alignment-like cells near the surfaces. The continuously twisted LC director distribution in the voltage-off state and tri-layered structure in the voltage-on state lead to several unique electro-optical characteristics of 90° TN cells between two crossed linear polarizers.

The normalized transmittance ($T_\perp$) of a TN cell with twist angle ϕ and cell gap d under crossed linear polarizers can be expressed as:

$$T_\perp = \cos^2 X + \left(\frac{\Gamma}{2X} \cos 2\beta \right)^2 \sin^2 X, \qquad (1.1)$$

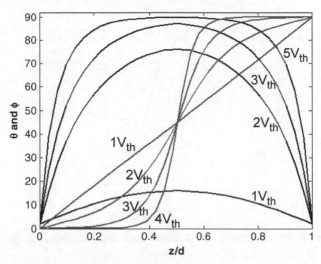

Figure 1.11 LC director profile of a TN cell at different voltages. Dark lines represent tilt angles and gray lines twist angles

where $X = \sqrt{\phi^2 + (\Gamma/2)^2}$, $\Gamma = 2\pi d\Delta n/\lambda$, and β is the angle between the polarization axis and the front LC director. As we can see, when $\cos^2 X = 1$ ($X = m\pi$ and m is an integer), the transmittance will be independent of β as the second term in the equation vanishes. By setting $X = m\pi$ and given $\Gamma = 2\pi d\Delta n/\lambda$, we can obtain the so-called Gooch–Tarry condition for a 90° TN cell ($\phi = 90°$):

$$\frac{d\Delta n}{\lambda} = \sqrt{m^2 - \frac{1}{4}}. \tag{1.2}$$

Typically, the first minimum condition with $m = 1$ and $d\Delta n/\lambda = \sqrt{3}/2$ is adopted for TN cells.

Figure 1.12 plots the voltage-dependent transmittance for three primary wavelengths of displays at R = 650, G = 550, and B = 450 nm. Since human eyes are most sensitive to green light, we optimize the transmittance of the TN cell at $\lambda = 550$ nm. In a 90° TN cell with β at 0° or LC rubbing direction parallel to the polarizer transmission axis, wavelength dispersion is relatively small (compared with a homogeneous or vertical alignment cell). In addition, the TN cell also has a relatively easy fabrication process and good light efficiency, making it the preferred technology in most LCD devices that do not require a wide viewing angle.

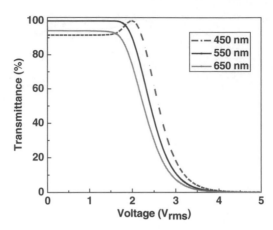

Figure 1.12 Voltage-dependent transmittance of a normally white 90° TN cell with $d\Delta n = 480$ nm

The drawbacks of a TN cell are its narrow viewing angle and gray level inversion when used without compensation film, resulting from its inherent LC director distribution in the dark state. Referring to Figure 1.11, the tri-layered structure in the dark state has two hybrid-alignment-like thin layers near the boundaries, and a vertical-alignment-like cell in the bulk. At normal incidence, the two thin boundary layers are crossed to each to cancel the phase retardation, leading to a good dark state. But at off-normal incidence, these three different layers all cause phase retardation, leading to a narrow viewing angle. The butterfly-shaped iso-contrast plot for the TN cell at $\lambda = 550$ nm with protective tri-acetyl cellulose (TAC) films is shown in Figure 1.13, where the contrast ratio drops quickly away from normal incidence. A relatively narrow viewing angle limits the application of the TN cell to small to medium-sized displays.

The narrow viewing angle results from several factors, e.g., optical anisotropy of LCs from the tri-layered structure in the dark state, off-axis light leakage from two crossed linear polarizers (or effective polarizer angle deviation), and light scattering in the color filters. To increase the viewing angle of a TN cell, compensation film is needed; this should be able to compensate for the tri-layered structure and the effective polarizer angle deviation. One commonly employed wide-view (WV) film is the discotic film developed by Fujifilm Company [26, 27]. Details of the compensation scheme are plotted in Figure 1.14, where a discotic material (triphenylene derivatives) is coated on an alignment layer on a TAC substrate near each polarizer's inner surface. The discotic material is in a hybrid alignment configuration with the

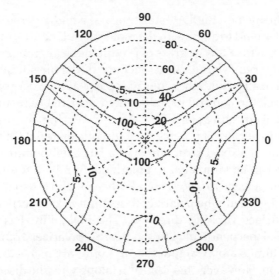

Figure 1.13 Iso-contrast plot of a 90° TN cell at $\lambda = 550$ nm with protective TAC films

following important features: (i) it has π-electrons spread in a disk-like shape, which gives rise to a high birefringence (much larger than a typical compensation film), and (ii) it exhibits a discotic nematic phase at a lower temperature than that at which the TAC starts to deform, enabling a uniform formation of the film on the TAC substrate [26]. The discotic material adjacent

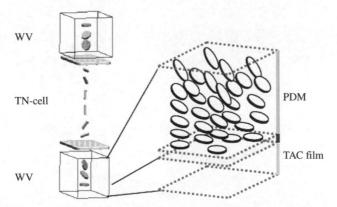

Figure 1.14 Wide-view discotic film for TN LCD (courtesy of K. Takeuchi *et al.* (27), Reproduced by permission of SID)

to the alignment layer on the TAC substrate has a high degree of randomness and its molecules tend to align with the molecular plane almost parallel to the alignment layer, with a few degrees of pre-tilt in the alignment layer rubbing direction. On the other hand, in the vicinity of the air surface, the discotic molecules tend to align with the molecular plane almost perpendicular to the air surface. Thus, the controllable alignment near the two surfaces forms a hybrid alignment structure of the film. After UV curing, the discotic material is polymerized to become a polymerized discotic material (PDM) like a film, which also exhibits a fixed structure over a large temperature range.

The WV film can be laminated to polarizing film by a roll-to-roll process. Optically, to compensate for the TN mode, the WV film is usually employed in an O-mode TN LCD with the polarizer transmission axis perpendicular to the adjacent LC surface rubbing direction. The PDM film has its azimuthal direction aligned perpendicular to the nearby polarizer transmission axis. Here the TAC film is a biaxial film (i.e., $n_z < n_y < $ or $\cong n_x$), exhibiting a small phase retardation. However, this TAC film (along with the discotic PDM film) is also critical in phase compensation. Figure 1.15 shows the iso-contrast plot of a TN LCD by incorporating the WV film with both PDM and TAC films. Compared with the previous result, the viewing angle is significantly

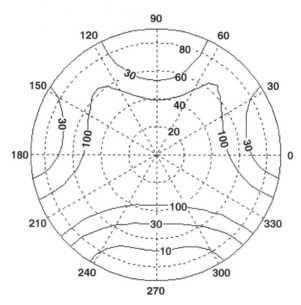

Figure 1.15 Iso-contrast plot of a 90° TN cell at $\lambda = 550\,\mathrm{nm}$ with both protective TAC films and WV films

increased and becomes more symmetrical. However, such viewing angle performance is still inadequate for applications like large-panel LCD TVs, for which advanced LCD modes like in-plane switching (IPS) and multi-domain vertical alignment (MVA) are more favored.

1.3.2 Homogeneous Alignment Mode

In a homogeneous alignment cell, the alignment layers are initially anti-parallel rubbed with a certain pre-tilt angle for the boundary LCs. When a relatively high voltage is applied between rear and front plane electrodes, vertical electric fields cause the LC molecules to tilt up. Figure 1.16 shows both the off-state and on-state LC director profiles and light polarization changes in a homogeneous cell between two crossed linear polarizers. The homogeneous alignment mode belongs to the ECB (electrically controlled birefringence) family, where a pure phase retardation effect dominates in modulating the polarization of the incident light. For a phase retarder (with thickness d, birefringence Δn, and wavelength λ) inserted between two crossed linear polarizers, if the optical axis is aligned at an angle of ϕ with respect to the absorption axis of the rear linear polarizer, the output light intensity can be written as $I = I_o \sin^2 2\phi \cdot \sin^2(\delta/2)$, where δ is the phase retardation value at $2\pi d\Delta n/\lambda$, and I_o is the maximum output light intensity from two such parallel linear polarizers. To obtain maximum transmittance, ϕ should be equal to $45°$ and $\delta = (2m + 1)\pi$, where m is an integer. In most cases, $m = 1$ in order to shorten the response time.

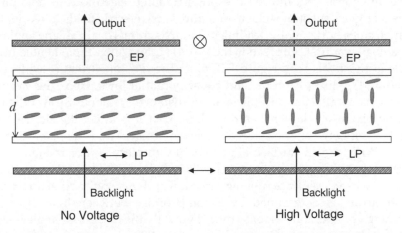

Figure 1.16 LC director distributions and changes of light polarization in a homogeneous cell in the voltage-off and voltage-on states

A homogeneous LCD with the LC rubbing direction at 45° to the transmission axis of the linear polarizer is a normally white display. Under a relatively high voltage, positive dielectric anisotropic ($+\Delta\varepsilon$) LCs will be driven to become vertically aligned to the substrate, to have a reduced effective δ and in turn lower the output transmittance. However, due to the strong surface anchoring force, surface LC molecules will still be aligned at their initial pre-tilt angles. Thus, a small residual wavelength retardation will exist even if a relatively high voltage is applied. In order to eliminate such surface residual retardation to achieve a good dark state, compensation films are needed. In applications using transflective LCDs, an ECB-based dual-cell-gap device is one of the mainstream technologies, where the surface phase residual retardation can be cancelled by quarter-wave plates in the adjoining circular polarizers, as will be discussed in Chapter 3. Generally, a homogeneous base LCD exhibits a narrow viewing angle, owing to the surface LC distribution in the dark state. In addition, with a single-domain structure, gray level inversion is also quite severe. To widen the viewing angle, complicated compensation films such as discotic films are also required.

1.3.3 In-plane Switching (IPS) Mode

In a TN or homogeneous cell, the LC molecules are perturbed by the longitudinal electric fields between the two substrates. Thus, different LC profiles and phase retardations are observed when viewed from the left, right, upper, or lower directions, yielding an asymmetrical and narrow viewing angle. In contrast, an IPS cell uses interdigitated electrodes formed on the same rear glass substrate to generate transverse electric fields to reorient the LC molecules horizontally, leading to a more symmetrical viewing angle. IPS technology was first proposed in the 1970s [11] and later implemented in TFT LCDs in the 1990s [12]. Figure 1.17 plots the IPS cell structure with corresponding LC director profile and equal potential lines. Here, the electrode width w is about 4 μm and the spacing distance l is about 8 μm. When the spacing distance is larger than the cell gap, strong horizontal electric fields mainly exist between the common electrode (black) and the pixel electrode (gray). With proper surface alignment, LC molecules are rotated by the horizontal fields to induce phase retardation of incident light. The optimized angle between the surface rubbing direction and the longitudinal direction of the electrode is determined by a compromise between light efficiency, operating voltage, and response time. Typically the optimized angle is about 10° when using a $+\Delta\varepsilon$ LC material and about 80° when using a $-\Delta\varepsilon$ one. On the other hand, as electric fields are mostly vertical in the regions in front of the

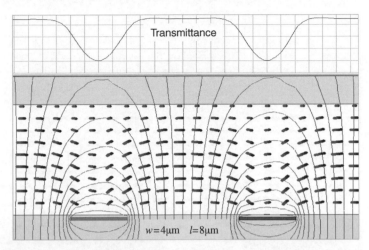

Figure 1.17 IPS cell configuration with director profile, equi-potential lines, and transmittance curve

electrode strips, LC molecules with a $+\Delta\varepsilon$ would tilt up most there, causing a loss of phase retardation and low transmittance for the incident light, as shown in the plot. Overall, the light efficiency in an IPS cell using a $+\Delta\varepsilon$ LC material is only about 75% of that of a TN LCD. Using a $-\Delta\varepsilon$ LC material in an IPS structure enhances the value to over 80%, but the driving voltage is increased because negative LC mixtures usually exhibit a smaller $|\Delta\varepsilon|$, i.e., a higher threshold voltage.

To improve light efficiency while maintaining the horizontal movement of LC molecules, fringe-field switching (FFS) mode was developed [13]. An FFS cell using a $-\Delta\varepsilon$ LC material is shown in Figure 1.18 with LC director profile and equi-potential lines included. Here, the common electrode is planar in shape and pixel electrodes are in strips. Typically the pixel electrode has a width $w \sim 3\,\mu m$ and a gap $g \sim 5\,\mu m$. Here, although the electrode gap g is larger than the cell gap (about $4\,\mu m$), the actual horizontal distance between a pixel electrode edge and the common electrode is zero. As a result, stronger fringe fields with both horizontal and vertical field components will be generated to perturb the LC molecules. Moreover, we can observe different dynamic mechanisms between an IPS cell and an FFS cell. In an IPS cell, strong horizontal electric fields are formed between the pixel and common electrodes to rotate the LC molecules parallel to the substrate, while the LCs in front of the electrode strips mainly tilt up. But in an FFS cell, fringe fields with much stronger horizontal fields are only formed near the electrode edges (equi-potential lines there are denser) and decrease towards the centers of electrode

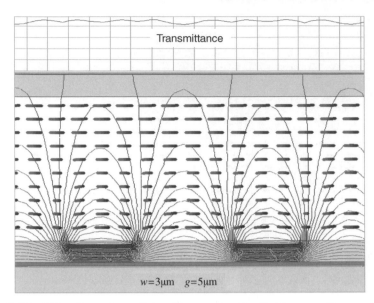

Figure 1.18 FFS cell configuration with equal potential lines and transmittance curve. The voltage difference between the front pixel electrode strips and the rear planar common electrode is 4.5 V_{rms}

strips or gaps. The strong horizontal fields near the edges rotate the LCs there in-plane first. Owing to the small dimensions and stronger fields, this perturbation near the edges in turn propagates horizontally in both directions to cause the reorientation of LCs near the centers of the electrodes or gaps where the field lines are vertical. Because of this unique process, the transmission of an FFS cell can be greatly enhanced to reach over 95% of that of a TN cell by using a $-\Delta\varepsilon$ LC material and over 85% by using a $+\Delta\varepsilon$ one. Usually, a $+\Delta\varepsilon$ LC material is preferable in an FFS LCD, since it has a larger $\Delta\varepsilon$ value and a lower viscosity, thus its operating voltage and response time can be reduced. In addition to the high transmission efficiency, other characteristics of an FFS cell make it more popular: (i) the capacitance between the front pixel electrodes and the rear common electrode can contribute to the storage capacitance of the pixel, which further improves the aperture ratio of the display, and (ii) because of the horizontal reorientation of LC molecules, an FFS cell exhibits weak color dispersion, which is an advantage compared to TN or VA cells. The second one is also true for IPS cells. More details of the FFS mode will be discussed in Chapter 4.

A detailed investigation of the mechanisms by which we can obtain high transmittance in IPS or FFS cells is of interest. To study the origins of high light

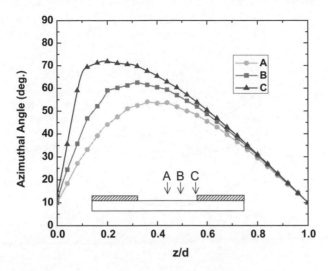

Figure 1.19 LC twist profiles at different positions in an FFS cell

efficiency, we plot, in Figure 1.19, the LC azimuthal angle distributions at different cell positions: gap center at position **A**, intermediate position **B** between gap center and electrode edge, and electrode edge at position **C**. The LC twist angle is largest at position **C** where horizontal fields are strongest, and decreases as the position approaches the center of the electrode gap. More importantly, the on-state LC profile is like two TN cells with opposite twist senses and the peak twist angle plays a critical role in the output light efficiency.

The polarization trace of light traversing the LC cells from these three different positions in an FFS cell at normal incidence is plotted in the S_1–S_2 plane (S_1 and S_2 are Stokes parameters), as shown in Figure 1.20. (For an introduction to light polarization in LCDs represented by Stokes parameters on the Poincaré sphere, the reader should refer to Chapter 3.) Light with linear polarization parallel to the rear polarizer transmission axis is defined as $S_1 = +1$ at point T_{rear}, which is the starting point for the light incident towards LC from positions **A**, **B**, and **C**. Thus, the front linear polarizer transmission axis perpendicular to the rear polarizer can be designated at point T_{front} by $S_1 = -1$. With a peak twist angle larger than 45° across a very thin cell gap at position **C**, the light polarization there quickly moves to the lower half-plane with $S_2 < 0$. And then the front TN structure rotates the polarization in a different curvature, adjusting the final polarization close to the point T_{front}. With a smaller peak twist angle (< 45° away from the initial angle) at position

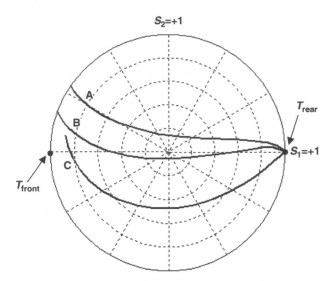

Figure 1.20 Polarization traces of light traversing the FFS cell from different positions at normal incidence

A, the light polarization trace is mainly above the S_1 axis in the upper half-plane with $S_2 > 0$. The inadequate maximum twist angle in both rear and front half TN cells makes the final polarization depart away from the point T_{front}, outputting a smaller light transmittance there. Therefore, peak twist angle is important for obtaining high transmittance, and a two-TN-cell structure would make the transmittance less sensitive to wavelength dispersion.

Figure 1.21 shows the VT curves of the FFS structure for three primary display wavelengths at R = 650, G = 550, and B = 450 nm. The driving voltage of this FFS cell using a $-\Delta\varepsilon$ LC material is about 4.5 V_{rms}, and the cell is optimized at a wavelength of 550 nm. An inset plot is also included to compare the VT curves at the three wavelengths, where each curve is normalized to its own maximum value. The R, G, and B curves overlap perfectly with each other, originating from the two-TN-cell configuration of the on-state LC director profile in the FFS cell, as analyzed earlier.

Another important feature of IPS and FFS cells is their inherent wide viewing angle. The iso-contrast plot for an uncompensated FFS cell using a $-\Delta\varepsilon$ LC material is plotted in Figure 1.22. A contrast ratio of 10:1 can be achieved to about 70° in all directions. If a $+\Delta\varepsilon$ LC material is used, the rotation of the LC directors will be influenced by the vertical electric field components, resulting in a lower transmittance ($\sim 85\%$) and a narrower viewing angle (contrast ratio 10:1 to about 60°). But a $+\Delta\varepsilon$ LC usually has a

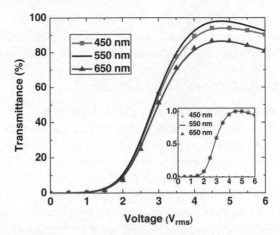

Figure 1.21 VT curves of R, G, and B light for a 4 μm normally black FFS cell. The inset plot shows the normalized curves; the three curves overlap perfectly

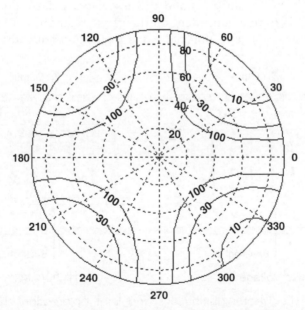

Figure 1.22 Iso-contrast plot of an FFS cell at $\lambda = 550$ nm without compensation films

larger $\Delta\varepsilon$ value and smaller rotation viscosity, leading to a lower driving voltage and faster response time. The wide viewing angle and weak color dispersion make FFS extremely attractive for small-panel mobile displays.

1.3.4 Vertical Alignment (VA) Mode

The homeotropic cell, also called a vertical alignment (VA) cell, is another mainstream LCD technology, owing to its super high contrast ratio. In a VA cell, the LC directors are almost perpendicular to the surface alignment layers, thus the axial light leakage between two crossed linear polarizers is much lower than in a homogeneous cell, TN cell, or even an IPS cell. Figure 1.23 shows the LC director profiles in the voltage-off and voltage-on states of a single-domain VA cell. In the voltage-off state, the incident light maintains its polarization when traversing the LC cell, and is then blocked by the front linear polarizer independent of the incidence wavelength, leading to a good dark state for all wavelengths. In a relatively high voltage state, the bulk LC directors tilt down, causing phase retardation for the incident beam. Similar to the homogeneous cell, for a single-domain VA cell, the optical axes of the surface LC molecules are controlled at $\pm 45°$ from the polarizer transmission axis for maximum transmission. The output transmission intensity I also satisfies $I = I_o \sin^2 2\phi \cdot \sin^2(\delta/2)$ and the associated polarization change of incident light is purely dependent on the phase retardation of the LC cell, which is similar to a pure uniaxial wave plate. Consequently, the output

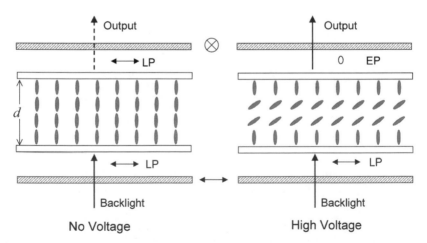

Figure 1.23 LC director distributions and light polarization changes in a homeotropic cell (or VA cell) in voltage-off and voltage-on states (LP: linear polarization; EP: elliptical polarization)

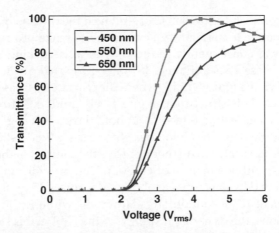

Figure 1.24 VT curves of R, G, and B wavelengths for a 4 μm VA cell

transmission is strongly wavelength dependent, which can be clearly seen from Figure 1.24.

A single-domain VA cell exhibits a narrow viewing angle and severe gray level inversion. To increase the viewing angle, both film compensation and a multi-domain structure are needed. Fujitsu [7–9] developed the first multi-domain vertical alignment (MVA) technology by introducing protrusions on both glass substrates, as shown in Figure 1.25(a). With small pre-tilt angles on the protrusion surfaces, the LC directors can be adjusted to tilt in different directions, forming two domains. A four-domain structure can be achieved with zigzag electrodes and protrusions. This early design required a complicated fabrication process to make the protrusions on both substrates, and

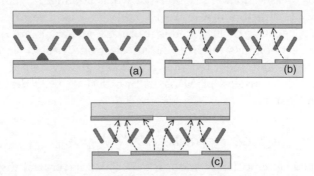

Figure 1.25 Structures of MVA cells using (a) protrusions only (by Fujitsu), (b) both protrusions and slits (by Fujitsu), and (c) slits only (PVA by Samsung)

exhibited low contrast due to the LC pre-tilt near the protrusions. To improve the contrast, opaque materials were employed to make protrusions, but the aperture ratio was consequently reduced by locally blocking light transmission. To solve these issues, Fujitsu [8] later proposed an improved MVA version, as shown in Figure 1.25(b), where the rear protrusions were replaced by electrode slits. This structure can effectively guide LC directors to form multi-domains by adjusting effects from both the pre-tilt angles on the front protrusions and the fringing fields near the rear electrode slits. Slits can be utilized on both substrates to further improve the contrast of the VA cell. This concept was first outlined in the work of Alan Lien and colleagues [28–30] and has been developed by Samsung, as shown in Figure 1.25(c), as the patterned vertical alignment (PVA) mode [16]. However, without the small pre-tilt on the protrusions, the dynamic response of the PVA cell is slower than the MVA cell using protrusions. To decrease the response time, special driving schemes to give the LCs an initial pre-tilt by means of a small bias voltage are helpful [31].

Owing to their high contrast and wide viewing angle (with compensation films), MVA and PVA LCDs are the technologies most widely employed for TV and monitor applications. For small-panel transflective LCDs, besides the wide viewing angle, MVA cells have the following features that are quite attractive: (i) the optical configurations for combining transmissive and reflective modes are relatively simple, requiring only circular polarizers, and (ii) it is possible to compensate for the optical path difference between transmissive and reflective modes by applying different electric fields to each mode in a single cell gap configuration. But VA mode also has certain drawbacks such as low response time, low light efficiency, and severe surface pooling. Surface pooling is an optical effect that occurs when the panel surface is touched or pressed, making it look like water pooling. It is caused by the distortion of LC directors inside the LC cell and a long restoration time is needed when negative dielectric anisotropic LC materials are used. Fortunately, new technologies such as novel electrode design and polymer-sustained surface alignment technology [32, 33] greatly help to solve these issues. Detailed discussion of MVA in transflective LCD designs will be presented in Chapter 4.

1.3.5 Hybrid Aligned Nematic (HAN) Mode

Hybrid aligned nematic (HAN) mode is a special liquid crystal alignment that combines different surface alignment on each of the two surfaces: a homogeneous alignment on one substrate and a vertical alignment on the other [34].

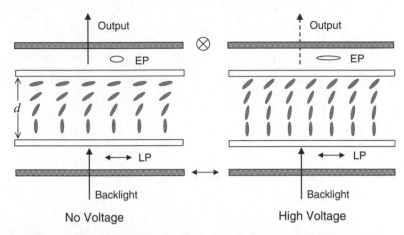

Figure 1.26 LC director distributions and changes of light polarization in a HAN cell in the voltage-off and voltage-on states

The LC director distributions of a HAN cell in the voltage-off and voltage-on states can be seen in Figure 1.26. Compared with a pure homogeneous cell or a VA cell with the same cell gap, the effective phase retardation of the HAN cell is reduced to about half of the $d\Delta n$ value. In addition, when a relatively high voltage is applied, the LC molecules near the homogeneous alignment surface remain parallel to the substrate, yielding residual phase retardation and light leakage.

This unique hybrid alignment structure enables several special display applications: (i) it is a useful alignment for the reflective region to achieve single-cell-gap operation in a transflective LCD, when the transmissive region uses a normal VA or homogeneous alignment, and (ii) it forms special optical compensation films to compensate for the dark state of the TN cell, homogeneous cell, or pi-cell that has a HAN profile near the substrate in the dark state [35, 36]. Detailed application examples will be discussed later.

1.3.6 Pi-cell or Optically Compensated Bend (OCB) Alignment Mode

The pi-cell, where the two substrate surfaces are treated with parallel rubbing, was first reported in 1984 [14, 15]. Typically the pi-cell is pre-controlled at an initial bend state, as shown in Figure 1.27(a). Without any treatment such as a bias voltage, the pi-cell usually stabilizes at a splay state. The transition from a splay state to a bend state takes quite a long time, sometimes even up to minutes for large panels. Thus, for fast switching purposes, it is preferable to

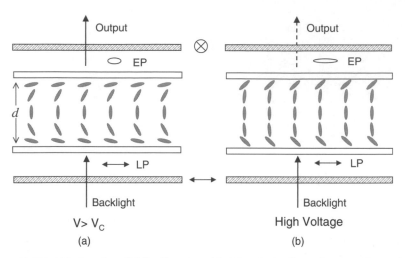

Figure 1.27 LC director distributions and light polarization changes in a pi-cell at (a) the initial state and (b) the voltage-on state

switch the cell between the bend state in Figure 1.27(a) and a high tilt state in Figure 1.27(b). In practice, to maintain a stable initial bend state, the LC free energy in the bend profile needs to be lower than that in a splay distribution. Approaches including a bias voltage above a critical voltage or high surface pre-tilt angles [37] can be utilized to stabilize the bend state.

The fast response time of the pi-cell, resulting from the flow effect and half-cell switching [14], is quite attractive. Figure 1.28(a) depicts the flow in a homogeneous cell with anti-parallel rubbing. After the electric field is switched off, the surface anchoring forces function to relax the LC molecules back to their initial state, causing horizontal flows in the cell. As indicated by the small arrows showing the flow directions, the LC directors around the center face a torque that rotates them opposite to the general tendency. This

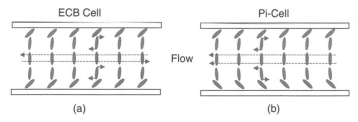

Figure 1.28 Comparisons of flows between (a) a homogeneous cell and (b) a pi-cell during the relaxation stage

causes 'back-flow' that slows down the switching-off time. Optically, the back-flow causes an 'optical bounce' during the relaxation process in the time-dependent transmittance curve. A pi-cell does not exhibit such a back-flow phenomenon, as shown in Figure 1.28(b). The flow directions originating from the front and rear surfaces have the same direction for the central LC directors. Thus, LCs near the bulk center behave in a 'stabilized' way and the switching-off time is much faster.

Similar to the homogeneous cell described above, in the voltage-on (or dark) state, residual phase retardation exists due to the surface LC molecular orientation. Thus, to obtain a common dark state for different incident wavelengths, compensation films are needed to cancel such residual phase retardation. In addition, in the dark state, the surface LC distribution is like that of a HAN cell, and a discotic film is also suitable for phase compensation of a pi-cell to suppress off-axis light leakage [38]. The fast response time and symmetrical viewing angle of the pi-cell make it quite attractive for mobile displays involving video applications.

1.4 Compensation Films

To obtain a high contrast ratio and wavelength-independent dark state, normally black modes like IPS (or FFS) and MVA (or PVA) are preferable to the normally white types using TN or homogeneous cells. In IPS (or FFS) and MVA (or PVA) cells, the LC layer in the voltage-off state does not modulate the polarization of normally incident light. But at oblique incidence, light leakage still occurs, especially in the bisector directions (the directions with an incident azimuthal angle ±45° away from the absorption axis of the polarizers). Such light leakage results from two sources: (i) the deviation of effective polarizer angle, and (ii) the off-axis phase retardation of the LC layer itself.

1.4.1 Deviation of Effective Polarizer Angle

Figure 1.29 shows a general representation of the effective angle between two auxiliary axes viewed on an oblique incident wave plane. In the x–y plane, for axis **OM** at azimuthal angle ϕ_1 and axis **ON** at angle ϕ_2 with respect to the reference x-axis, their angular difference is $\phi_2 - \phi_1$ when viewed from the normal z-axis. But this angular difference varies as the viewing direction changes from the z-axis to an oblique direction along **OK** in the plot. Based on the dot product of vectors, the effective angle $\angle$MKN between the projected lines **KM** and **KN** viewed on the wave plane **MKN** can be calculated and expressed as [23]:

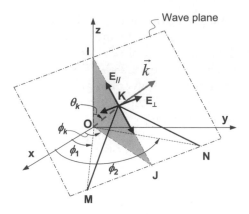

Figure 1.29 A representation of the effective angle viewed on an oblique incident wave plane between two auxiliary axes

$$\varphi = \cos^{-1}\left[\frac{\cos(\phi_2-\phi_1)-\sin^2\theta_k \cos(\phi_2-\phi_k)\cos(\phi_1-\phi_k)}{\sqrt{1-\sin^2\theta_k \cos^2(\phi_2-\phi_k)}\sqrt{1-\sin^2\theta_k \cos^2(\phi_1-\phi_k)}}\right], \qquad (1.3)$$

where θ_k and ϕ_k are the polar angle and azimuthal angle in the medium of the wave vector $\vec{k}$ respectively. For light incident at a polar angle of θ_0 from the air on to the medium, the polar angle may be obtained as $\theta_k = \sin^{-1}(\sin\theta_0/n_p)$ in the medium with n_p (~1.5) defined as the real part of the average refractive index of the medium.

For the special case of two crossed linear polarizers with $\phi_2 - \phi_1 = 90°$, the above equation can be written in another form in terms of $\phi_1 - \phi_k$ as:

$$\begin{aligned}\varphi &= \cos^{-1}\left[\frac{1/2 \cdot \sin^2\theta_k \sin 2(\phi_1-\phi_k)}{\sqrt{1-\sin^2\theta_k + (1/2 \cdot \sin^2\theta_k \sin 2(\phi_1-\phi_k))^2}}\right] \\ &= \cos^{-1}\left[\frac{sign[\sin 2(\phi_1-\phi_k)]}{\sqrt{(1-\sin^2\theta_k)/(1/2 \cdot \sin^2\theta_k \sin 2(\phi_1-\phi_k))^2 + 1}}\right].\end{aligned} \qquad (1.4)$$

Therefore, at any polar angle of θ_0, when $\phi_1 - \phi_k = \pm 45°$, the absolute value of $(\varphi - 90°)$ reaches the maximum at $\pi/2 - \cos^{-1}\left(\frac{\sin^2\theta_0/n_p^2}{2-\sin^2\theta_0/n_p^2}\right)$, i.e., the effective angle of two polarizers deviates the most from the initial crossed configuration

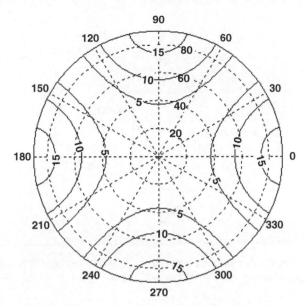

Figure 1.30 Angular-dependent deviations of the effective polarizer angle from 90°

at 90°. Thus, the incident azimuthal directions at ±45° from the polarizer transmission axis are defined as the bisector directions, where light leakage is most severe. Figure 1.30 plots the deviation of the effective polarizer angle from 90° ($\varphi - 90°$ in absolute value) with respect to different viewing azimuthal angle values ($\phi_1 - \phi_k$), which agrees well with the above analysis. For two crossed linear polarizers, normally incident light could be well absorbed, as the light passing the rear linear polarizer has a polarization direction parallel to the absorption axis of the front one. However, at off-axis incidence (such as at a polar angle of 70° along the bisector direction), the two polarizers are no longer perpendicular to each other and the light passing the rear linear polarizer will have a polarization direction away from the front linear polarizer absorption axis, leading to light leakage. To suppress the light leakage resulting from this deviation of the effective polarizer angle, compensation films can be used to adjust the light coming out of the rear linear polarizer to coincide with the front linear absorption axis. Detailed analysis of the compensation method based on the Poincaré sphere will be discussed in Chapter 3.

1.4.2 Phase Retardation from Uniaxial Medium

Besides the deviation of effective polarizer angle, the phase retardation of the LC cell itself also causes a change in polarization and leads to imperfect light

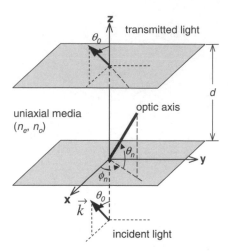

Figure 1.31 Schematic view of light propagation through a uniaxial medium with the incident wave vector in the x–z plane

absorption. The LC layer can be treated as a uniaxial layer with extraordinary and ordinary refractive indices n_e and n_o. When light propagates through a uniaxial medium, generally two eigenwaves with different phase velocities are excited, thus phase retardation occurs, which changes the light polarization. Figure 1.31 depicts light propagation through a uniaxial medium. Without losing generality, the optical alignment of incident light can be set into the x–z plane with a propagation vector $\vec{k}$ of (sin θ_0, 0, cos θ_0), and a uniaxial layer with extraordinary and ordinary refractive indices at n_e and n_o, optical axis at (cos ϕ_n, sin θ_n), and a thickness of d.

The phase retardation of a uniaxial medium at oblique incidence can be expressed by the phase difference between the extraordinary wave, which has a wave number k_{ez} along the z-axis, and the ordinary wave, which has k_{oz} as $\Gamma = (k_{ez}-k_{oz})d$ [39–41]. A solution of the Maxwell equation for wave propagation through a uniaxial medium can lead to expressions for k_{ez} and k_{oz} as follows [39–41]:

$$
\begin{aligned}
k_{ez} &= k_0 \left[-\frac{\varepsilon_{xz}}{\varepsilon_{zz}}\frac{k_x}{k_0} + \frac{n_o n_e}{\varepsilon_{zz}} \sqrt{\varepsilon_{zz} - \left(1 - \frac{n_e^2 - n_o^2}{n_e^2}\cos^2\theta_n \sin^2\phi_n\right)\left(\frac{k_x}{k_0}\right)} \right] \\
&= \frac{2\pi}{\lambda} \left[\frac{n_e n_o}{\varepsilon_{zz}} \sqrt{\varepsilon_{zz} - \left(1 - \frac{n_e^2 - n_o^2}{n_e^2}\cos^2\theta_n \sin^2\phi_n\right)\sin^2\theta_0} - \frac{\varepsilon_{xz}}{\varepsilon_{zz}}\sin\theta_0 \right],
\end{aligned}
\tag{1.5}
$$

and

$$k_{oz} = k_0 \left[\sqrt{n_o^2 - \left(\frac{k_x}{k_0}\right)^2} \right] = \frac{2\pi}{\lambda} \left[\sqrt{n_o^2 - \sin^2\theta_0} \right]. \tag{1.6}$$

where λ is the incident wavelength, ε_{zz} and ε_{xz} are the tensor components of the uniaxial medium, $\varepsilon_{zz} = n_o^2 + (n_e^2 - n_o^2)\sin^2\theta_n$ and $\varepsilon_{xz} = \varepsilon_{zx} = (n_e^2 - n_o^2)$ $\sin\theta_n \cos\theta_n \cos\phi_n$. Detailed derivation of these equations will be presented in Chapter 2. With these known variables, the phase retardation $\Gamma = (k_{ez} - k_{oz})d$ of the light propagating through the uniaxial layer can be determined.

For a homogeneous LC cell (if the small surface pre-tilt angle is neglected), the LC layer behaves optically like a positive uniaxial A-film; for a vertical alignment LC cell, the LC layer behaves optically like a positive uniaxial C-film. With $\theta_n = 0°$ and $90°$ for the A-film and C-film respectively, the phase retardations can be expressed as:

$$\Gamma_a = \frac{2\pi}{\lambda} d \left[n_e \sqrt{1 - \frac{\sin^2\theta_0 \sin^2\phi_n}{n_e^2} - \frac{\sin^2\theta_0 \cos^2\phi_n}{n_o^2}} - n_o \sqrt{1 - \frac{\sin^2\theta_0}{n_o^2}} \right], \tag{1.7}$$

and

$$\Gamma_c = \frac{2\pi}{\lambda} d \left[n_o \sqrt{1 - \frac{\sin^2\theta_0}{n_e^2}} - n_o \sqrt{1 - \frac{\sin^2\theta_0}{n_o^2}} \right]. \tag{1.8}$$

1.4.3 Uniaxial and Biaxial Films

From the above analysis, the off-axis phase retardation of the LC cell could change the incident light polarization, causing light leakage. To minimize off-axis light leakage, compensation films are needed to operate on light leaving the LC layer, so that when it reaches the second linear polarizer it is linearly polarized parallel to the absorption axis of that polarizer. Figure 1.32 shows an optical retardation film with refractive indices n_x, n_y, and n_z and of thickness d. Optical compensation films can be divided into different categories based on different standards. For example, based on the relation between its refractive indices n_x, n_y, and n_z, they can be classified as uniaxial films with only one optical axis or biaxial films with two optical axes. Uniaxial films can further be classified into A-films and C-films, where an A-film has its optical axis parallel to the film surface ($n_e = n_x \neq n_o = n_y = n_z$) and a C-film has the optical

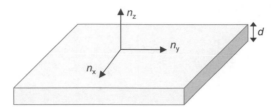

Figure 1.32 Typical parameters of a phase compensation film

axis perpendicular to the film surface ($n_e = n_z \neq n_o = n_x = n_y$). Furthermore, both A-films and C-films can be divided into positive or negative ones, depending on the relative value between the extraordinary refractive index n_e and the ordinary refractive index n_o: a positive film has $n_e > n_o$ and a negative one has $n_e < n_o$. On the other hand, if $n_x \neq n_y \neq n_z$, the retardation film has two optical axes, and is called a biaxial film. In addition to directly using n_x, n_y, n_z and its thickness d, a retardation film is also widely characterized in other ways by:

1. N_z factor, where $N_z = \frac{(n_x - n_z)}{(n_x - n_y)}$;
2. in-plane phase retardation $R_0 = (n_x - n_y)d$;
3. in-thickness phase retardation $R_{th} = [(n_x + n_y)/2 - n_z] \cdot d$;
4. average refractive index.

These parameters represent different optical interpretations of the phase retardation [42]. These uniform films all have their optical axes aligned in the same direction, either parallel or perpendicular to the film surface. In addition, there are some special films that have their optical axes inclined or non-uniformly distributed. For example, the above-mentioned Fuji film has negative discotic molecules in a hybrid alignment, an NH film has regular nematic LC molecules in a hybrid alignment, and an O-plate has its nematic LC molecules uniformly inclined by a certain tilt angle [43]. These films are mainly used to compensate LCDs having irregular LC director profiles in the dark state, such as TN cells, homogeneous cells, and pi-cells.

Optical films can be formed by different methods such as coating liquid crystalline materials or stretching polymer films [17, 44, 45]. For example, an A-film can be formed by coating and polymerizing discotic LC molecules in a sideward alignment for negative birefringence, like coins standing on edge, or by coating and polymerizing nematic rod-like LC molecules in a homoge-neous alignment for the positive type, or by stretching polymer materials to align polymer chains with the stretching direction. Whether this creates positive or negative birefringence depends on the side polymer chain structure and its alignment to the backbone. A C-film can be formed by coating and polymerizing nematic rod-like LC molecules in a homeotropic alignment for positive birefringence or short-pitched cholesteric nematic LC molecules for negative birefringence. Fuji films or NH films can be formed by coating discotic material or nematic material on a special alignment layer whilst controlling the formation conditions such as temperature. Biaxial films can also be formed by using the polymer-stretching method by stretching in two directions or by the coating method. Presently, fabrication methods for most uniaxial films are well developed, except for negative A-films, for which it is still difficult to obtain large birefringence and good quality in mass production. The fabrication technology for biaxial films has advanced quickly, and now these films are widely used for IPS and MVA LCDs, for which only uniaxial films dominated previously. However, it is still difficult to make biaxial films with certain pre-designed refractive indices and thickness, but sometimes special biaxial film parameters can be realized by stacking two or more different attainable biaxial films to produce the required optical performance.

1.5 Reflectors

The reflector and transflector are important components of transflective LCDs, since dual functions (transmissive and reflective) exist in the device simultaneously. An ideal reflector or transflector is required to reflect the incident ambient light towards the viewer in a confined cone with high reflectivity but also with a certain amount of diffusion to widen the viewing angle; and it should also avoid overlap between the major reflected beams and the front surface specular directional ambient light reflection. Surface treatment is critical to meet these requirements. In addition, the position of a reflector or transflector is also critical in order to avoid parallax. To better understand these requirements, the two concepts of parallax and ambient contrast ratio will be introduced. Then, different types of reflectors and transflectors and their application to mobile LCDs will be addressed.

1.5.1 Parallax and Ambient Contrast Ratio

Parallax occurs when a reflective LCD is viewed at an oblique angle [5], when the directly viewed image of a pixel and its image viewed in the reflector are separated by a distance comparable to or greater than the pixel size. It causes cross talk between pixels and loss of contrast. Figure 1.33 demonstrates an example of parallax in a reflective LCD when the polarizer and reflector are

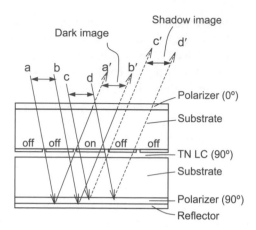

Figure 1.33 Schematic illustration of the parallax phenomenon in a reflective LCD

laminated on the outer side of the rear thick substrate. This sample is a normally white 90° TN reflective LCD that is widely used in low-end displays like electronic watches or calculators. At normal incidence, when no voltage is applied (off-state), the incident ambient light is transmitted by the LC cell with a 90° polarization rotation to pass through the rear linear polarizer and is then reflected, with an opposite 90° rotation, to pass back through the front linear polarizer to produce a bright image. When a relatively high voltage is applied (on-state), the incident light passing the LC cell sees a negligible rotation and is then blocked by the rear linear polarizer, leading to a dark image. When looking at the on-pixel at an off-axis direction, as shown in Figure 1.33, the viewer sees two types of 'dark' image. The first one, designated **a′–b′**, originates from the incident light source **a–b**, where the light first traverses the left off-pixel (having a 90° rotation) and further through the rear linear polarizer on to the reflector, and is then reflected back to transmit the on-pixel and finally becomes absorbed by the front linear polarizer. The second dark image, designated **c′–d′**, originates from the incident light source **c–d**, which passes through the on-pixel (maintaining its polarization) and is finally blocked by the rear linear polarizer before reflection, and **c′–d′** is its image to the viewer. Here, dark image **a′–b′** is caused by the front linear polarizer and image **c′–d′** results from the rear linear polarizer. The thick rear substrate and linear polarizer make the incident and exit light pass different pixels, creating cross talk between pixels. Images **a′–b′** and **c′–d′** shift from each other; this phenomenon is called *parallax*. The image

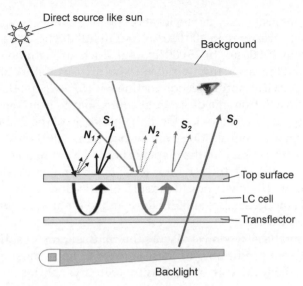

Figure 1.34 Consideration of sunlight readability of a reflective/transflective LCD

overlap results in deterioration of the observed image. To avoid such a parallax problem, the reflector or transflector is required to be formed close to the image pixel.

Ambient contrast ratio is another important concept that needs to be discussed to help us to understand the reflector/transflector design considerations. Figure 1.34 shows a reflective or transflective LCD under ambient lighting. There are two major ambient light sources: the direct light source such as collimated sunlight impinging on the display in a fixed direction, and the light from the background such as the blue sky and the multiple reflections or scattering of sunlight from various objects, which impinges on the display from every direction. From this illustration, two types of major noise source are observed: reflection N_1 from the collimated direct sunlight (assuming the surface is *not* an anti-glare surface), and reflection N_2 from the background light. Here, the reflection N_1 is confined mainly to the specular direction but small amounts of reflection occur at angles away from the exact specular direction. On the other hand, the light for reflection N_2 comes from all directions, thus N_2 may be viewed as uniformly distributed noise. Besides these surface reflections, another type of noise may come from stray ambient light that reaches the eye directly from scattering elsewhere, which is not shown in the figure but which we designate N_0. The direct sunlight and background intensity vary as the location or time change. Typically, direct sunlight can be greater than 100 000 lux and in full daylight the background may vary from 10 000 lux to 25 000 lux [46]. Such strong ambient light can easily wash out the image from a transmissive display (with a 200-nit surface luminance), even if an anti-reflection coating with $SR < 1\%$ is adopted. For the signals, we have S_1 from the direct light source and S_2 from the background after traversing the LC cell twice. Or, if the backlight of the display is turned on, another signal S_0 contributing from the backlight will also exist. Therefore, the ambient contrast ratio of the display should be defined as $ACR = (S_0 + S_1 + S_2)/(N_1 + N_2) + 1$ (please refer to Chapter 6 for more details of this ACR derivation). However, not all the surface reflections of the ambient light will be within the acceptance angle of the eye, and the above equation is just a qualitative expression.

From this analysis, when we discuss the contrast ratio of a display under complex ambient conditions, all these sources of noise must be taken into account. Detailed analysis of display readability will be discussed in a later chapter.

1.5.2 Reflector Designs

The light source and reflection pattern discussed above give rise to some requirements for designing and optimizing a reflective or transflective LCD. Typically, the human eye is located about 30 to 50 cm from the panel surface and is axially within $10°$ of the surface normal direction. As a result, for high brightness and a wide viewing angle, an ideal reflector or transflector should be able to collect most random incident light and redirect it out to the viewer in a confined cone centered about $10°$ from the normal direction. For such a case, a reflector with a bumpy surface is preferred. In addition, for good contrast ratio, it should also have the ability to deflect the output light S_1 away from the specular reflection N_1, i.e., the angles of incidence and exit need to be asymmetrical.

A conventional reflector widely used in reflective and transflective LCDs is a diffusive type, in which bumpy structures are formed directly on the flat substrate surface. A dielectric layer is first coated on to a flat substrate, then it is patterned by photolithography to make random bumps (randomness needs to be controlled), and finally a metal layer is coated on to the bumpy surface. This type of bumpy reflector can diffusively reflect the incident light into a wide viewing cone that is much wider than the useful $10°$ cone for human eyes. Thus, a large amount of incident light is wasted, yielding insufficient useful reflectance and contrast ratio. An alternative method to gain both satisfactory contrast and brightness is to form a diffusive micro

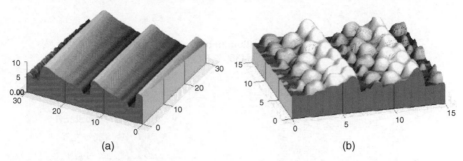

(a)   (b)

Figure 1.35 (a) A micro slant reflector (MSR) and (b) a diffusive micro slant reflector (DMSR) with a micro bumpy diffusive surface (courtesy of (43) and (44), Reproduced by permission of SID)

slant reflector (DMSR) or transflector, as shown in Figure 1.35 [47, 48]. The DMSR in Figure 1.35(b) can use its bumpy surface to collect random incident light and deflect the output beam away from the major specular reflection direction into the useful viewing cone. Compared with a conventional flat reflector structure, fabricating a DMSR requires more photolithographic steps. For example, first, the photo-resist is exposed by a multi-step exposure method to create the MSR structure in Figure 1.35(a), then a second exposure is used to obtain the bumpy surface profile, and finally a metal such as aluminum is deposited on the surface. The diffusive property is highly related to the second exposure step which controls the height and pitch of the bumps. To cover the entire visible spectrum, multiple bump pitches are required.

In addition to DMSR structures, researchers [49] have recently proposed utilizing multiple (~25) surface relief micro-gratings combined into one reflector to effectively collect more light from every direction; the diffractive reflector structure and its output pattern are shown in Figure 1.36. As we can see, gratings with different pitch lengths and inclined angles are used to cover the wavelengths in the visible range and to diffract the output light into a cone that is centered on the normal direction. As a result, the major output deviates from the specular reflection and more light is confined to the useful viewing cone. From these simulated results, this structure could improve the reflectance in the normal direction by a factor of 2 compared with the conventional flat bumpy reflectors. However, fabrication might be somewhat complicated.

Another type of reflector of interest is the nano wire grid polarizer (WGP) [50–52]. Figure 1.37 shows the schematic structure of a nano wire

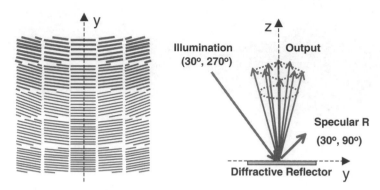

Figure 1.36 Structure and reflective pattern of the diffractive reflector (redrawn from (49))

grid polarizer having metal strips periodically formed on a glass substrate. When the pitch length P is far shorter than the incident wavelength, this metal grating structure functions like a polarization-dependent reflector. For unpolarized light incident on the WGP surface, the wave components with polarization parallel to the metal ribs will excite the unrestricted movement of electrons. The electron movement, in turn, excites a forward traveling wave as well as a backward or reflected wave, with the forward traveling wave canceling exactly the incident wave in the forward direction in the same polarization, yielding a strong reflection for this polarization. In contrast, if the incident wave is polarized perpendicular to the wire grid, the movement of electrons along this direction is restricted and the incident wave has substantial transmission. According to these properties, a wire grid

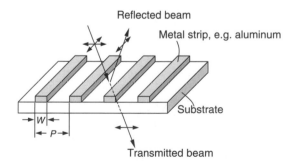

Figure 1.37 Schematic illustration of a nano wire grid polarizer and its characteristics

polarizer can be viewed as a polarization-dependent reflector for reflective and transflective LCDs or a reflective polarizer for the application of backlight recycling. Detailed discussion of the application of nano WGPs in displays will be addressed in Chapter 4.

1.6 Backlight

1.6.1 Backlight Configuration

The backlight configuration is different for a small-panel mobile phone and a large LCD TV. For mobile displays, edge lighting using LEDs is most desirable to achieve the required slim profile. A typical backlight configuration for a mobile display is shown in Figure 1.38(a), where light from edge LEDs is coupled into the light guide plate (LGP) which has patterns on its rear surface enabling light to be scattered towards the LCD cell. A non-metallic reflector with high reflectivity (>95%) is laminated behind the LGP to reflect any light leaving the LGP in a direction away from the LCD. To remove the pattern on the rear LGP surface (which creates hot spots) from the reflected light and make the output more uniform, a diffuser foil is then laminated on the front surface of the LGP. Further, one (oriented in the X or Y direction) or two (one oriented in each of the X and Y directions) prism films such as 3M´s BEF [6] are then attached to collect and collimate off-axis light into a cone within about $\pm 35°$ of the normal axial direction. In front of the prism films, there could either be no diffuser, for thickness and cost considerations, or a second diffuser to protect the prism films and reduce the moiré between the periodic structure of the films and pixel arrays. For backlight recycling, a reflective polarizer such as 3M´s DBEF (dual brightness enhancement film) [53] or a wire grid polarizer sheet [50–52] could also be placed before the rear sheet linear polarizer. The function of each different backlight element is illustrated in Figure 1.38(b).

For mobile display applications, low cost, minimum thickness, and high power efficiency from the backlight unit are most critical. To obtain good light efficiency for high brightness, each element, such as the LGP´s extraction factor, needs to be optimized. To reduce the thickness, researchers and engineers attempt to combine one or two prisms, and even the rear diffuser into the LGP structure, but a drawback is fabrication complexity, resulting in a much higher cost. A great deal of optimization work in both optical design and fabrication processes still needs to be conducted in order to perfect these mobile display devices.

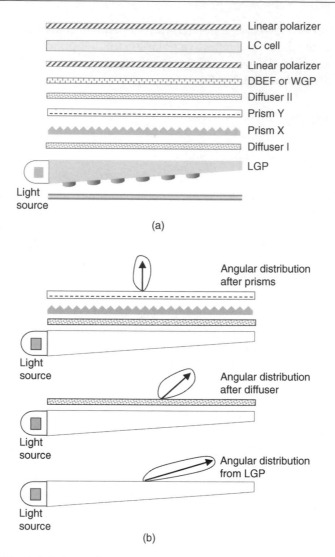

Figure 1.38 (a) Schematic configuration of the backlight unit in a mobile display; (b) angular light profile after each element

1.6.2 CCFL and LED Light Sources

The cold cathode fluorescent lamp (CCFL) is extensively used as a light source for LCDs owing to its ability to produce very bright white light at low cost and high efficiency. The CCFL is a discharge lamp constructed from a phosphor-coated glass cylinder with a mixture of mercury and a gas such as argon at low pressure sealed inside it with an electrode at either end. When a relatively high voltage is applied to the anode and cathode in the CCFL, a strong electric field is generated to accelerate electrons between the two electrodes. The collision of electrons with the mercury vapor excites mercury atoms above the ground state. Adding a fill gas enhances ionization of the mercury vapor at a lower voltage. Ultraviolet energy is released as excited mercury atoms decay back to the stable ground state, which in turn stimulates the phosphor to emit light in the visible spectrum. To generate white light for display purposes, tri-phosphors with individual red, green, and blue emission are utilized in the phosphor coating. The ratio of these different phosphors determines the output spectrum of the white light.

A typical spectrum of CCFL light is shown in Figure 1.39. As a reference, the spectra of the R, G, and B color filters used in LCDs are also included as dashed lines. The three primary peaks of the CCFL spectrum are located near 435 nm, 545 nm, and 615 nm. The green peak exhibits the narrowest FWHM (full width at half maximum) peak compared with the blue and red ones. However, due to the properties of the phosphors, there are still two undesired secondary peaks located near 490 nm and 585 nm, resulting in poor color

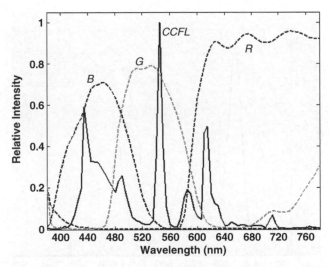

Figure 1.39 Emission spectrum of a CCFL light source (solid line) and the transmission spectra of RGB color filters (dashed lines). Note: the intensity is on a relative scale

separation and a smaller color gamut of about 72% of NTSC (National Television System Committee) in the CIE (International Committee on Illumination) 1931 plot. By shifting both primary and side peaks towards the red and adjusting the relative intensity of each peak by modifying the phosphors to better match the color filter spectrum, more highly saturated green and red colors can be achieved, improving the color extent to about 92% of NTSC [54]. Presently, CCFLs are still widely used for medium-sized and large LCD panels, because of their low cost, good stability, and simple thermal management.

LEDs are another widely used backlight source for LCDs (dominating in small display applications such as cell phones) owing to their compact size and steadily improving brightness. To obtain white light using LEDs, three different methods can be adopted [55]. One method is to mix light from independent R, G, and B LEDs. This method provides a very large color gamut, and is inherently the most efficient. The spectrum obtained by this method is shown in Figure 1.40. Using RGB LEDs enables local dimming to be used, enhancing the dynamic contrast ratio (to $>10^6$:1 under dark room conditions) and reducing power consumption (for DCR characterization, the backlight is underpowered in the dark state, while for a regular static contrast ratio measurement, the backlight is still fully on in the dark state). However, to

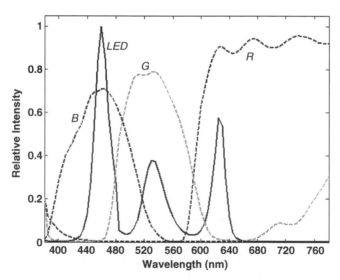

Figure 1.40 Emission spectrum of a white light source using R, G, and B LEDs

accurately stabilize the white point and reduce color shift originating from temperature variation and differential aging of the three colors, complicated driving electronics are needed. In addition, color mixing is not easy, especially in small-panel mobile displays, since the available coupling distance for different colored light is limited. Thus, this method is only used in high-end medium to large LCD panels such as more expensive notebooks and LCD TVs.

A second approach for achieving white light is to use a UV LED to pump R, G, and B phosphors. This method exhibits high color rendering, a stable white point, and a simpler driving circuit. But the efficiency is not good due to the inherent properties of the phosphors. In addition, the variation of the white point with polar angle is large. Moreover, with the existence of the UV LED, more consideration has to be given to the packaging of the backlight in order to avoid degradation from the UV light of other optical elements [55].

A third approach, and the dominant one for small panels, is to use a blue LED (InGaN) to pump a yellow phosphor (YAG). This design converts part of the blue LED light efficiently to a broad spectrum centered at about 580 nm, thus the combination of blue and yellow gives a pseudo white color. The spectrum of this white LED is shown in Figure 1.41. This method is simple and

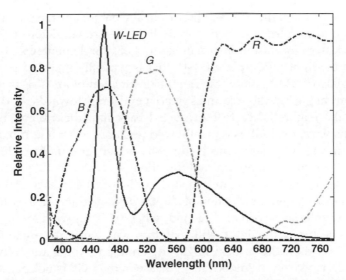

Figure 1.41 Emission spectrum of a white light source using blue LED-pumped yellow phosphor

provides good color rendering, while exhibiting excellent stability and reliability. As with the second approach, the efficiency is limited due to the phosphors. In addition, in practice, both the peak wavelengths of the emissive blue LEDs and the thicknesses of the yellow phosphors have certain variations, resulting in variations in the output white points. Thus, LEDs have to be binned according to their spectral characteristics for users to choose, and manufacturing variations arise as a result of these factors.

When using a white light source, the spectrum must be evaluated in accordance with the transmission spectra of the color filters used in LCDs. To obtain a large color gamut, good separation of the three primary colors is important. Referring to Figures 1.40 and 1.41, the spectrum using three R, G, and B LEDs exhibits the best color separation with the largest color gamut. For the CCFL and W-LED, a smaller color gamut is achievable owing to poor color separation. To improve optical performance characteristics such as luminous efficacy and color gamut, the spectra of the light source and color filters need to be further improved to increase primary color separation and improve green light components to overlap the human eye sensitivity function, centered at about 555 nm.

1.6.3 Other Backlight Elements and Films

In addition to the light source, other optical elements in the backlight unit also play critical roles in directing more light towards the viewer. Referring to the optical configuration shown in Figure 1.38, the element next to the light source is the light guide plate (LGP) that is typically made of polymethyl methacrylate (PMMA) material. The major function of an LGP is to couple out light from the light source and spread it uniformly towards the display or towards the rear reflector to be recycled back into the system. For small mobile displays, the LGP is typically wedge shaped with micro patterns on its rear surface. Optically, the wedge shape will gradually collimate the incident beam towards the normal direction. As a simple illustration, for a wedge-shaped LGP with a smooth surface (Figure 1.42), each reflection on one interface will give a reduction of β in the next incident angle with respect to the new interface normal direction, where β is the wedge angle. After several reflections, the light will exit the LGP, once the incident angle becomes less than the critical angle (determined by the refractive index difference). However, a smooth rear surface of the LGP is not preferred. As we can see, it is possible that some light will exit the LGP through the rear surface, and will be reflected by the reflective sheet. Also, some light will

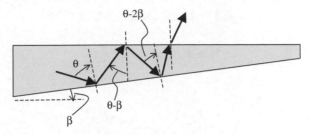

Figure 1.42 Light path in a wedge-shaped LGP with a smooth surface

travel to the right-hand end of the LGP and will be lost. Besides, the light output intensity distribution along the display will not be uniform with a smooth rear surface. To reduce light loss and obscure the output light (to eliminate any possible fixed patterns from the backlight structures before light exits the LGP front surface), certain microstructure patterns need to be formed on the rear LGP surface (as shown in the backlight configuration in Figure 1.38) to scatter (or on the front surface to extract) the incident rays. For low cost, a molding technique with the addition of a chemical etching step or a microstructure method can be employed to form such micro-patterns on the rear LGP surface. The uniformity can be adjusted by the pattern distribution and density. In all, by adjusting the micro-patterns on the LGP surface (dots are sparse near the light source and become denser as they approach the other end), the final output from the LGP will be suitable for further improvement by the optical elements between the LGP and the LCD.

In front of the LGP is a diffuser plate that functions as a means of eliminating images of the rear patterns of the LGP from the reflected light and making the output light more uniform in both a spatial and an angular sense. Typically, several mechanisms and methods are used to diffuse light, as shown in Figure 1.43 [56]. The bulk diffusion method uses a plate made of a bulk material mixed with tiny particles with a small refractive index difference between these two materials. Such a diffuser cannot collimate light. Another method is to engineer non-uniform textures on the exit surface of the diffuser: light collimation can be controlled by forming a particular surface slope distribution. Similarly, spherical beads can also be coated in a random distribution on the exit surface to generate diffusion and control the collimation of the light at the same time. Another interesting method involves the formation of random micro-lenses on a surface as the diffuser, so when the focal length of each micro-lens is small, the exit light from the

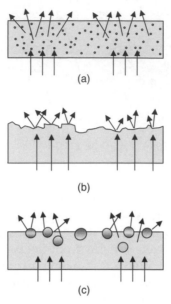

Figure 1.43 Different types of diffusion: (a) bulk diffusion; (b) engineered surface texture; and (c) surface beads (redrawn from (56))

micro-lens will diverge in different directions, creating an effect similar to a diffuser.

After the diffuser plate come the prism films designed to collimate the incident light. A 90° prism structure is shown in Figure 1.44, where the prism

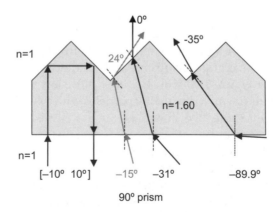

Figure 1.44 Schematic structure of a 90° prism sheet with some typical ray traces

refractive index is 1.60. Therefore the critical angle for an internal reflection is about 38.7°. Several representative rays are shown in the plot to illustrate the prism´s working mechanism. When the incident ray has an angle within 10° of the normal axial direction, the beam will be totally internally reflected back and can be recycled by being bounced back from the reflective sheet and being diffused by other elements. When the incident angle increases, the output ray is transmitted and is refracted over a certain angular range. Besides these typical rays, there are other rays not shown in the plot, for which the optical path is more complicated. In addition, a single prism sheet only confines light in one dimension. A second prism sheet placed perpendicular to the first is needed to collimate the output beam into a cone along the axial direction. As an illustration, the simulated overall light output patterns from the rear LGP and the following two prisms are plotted in Figure 1.45.

Because of the periodic structure of the prism sheet, a moiré pattern, defined as an optical effect resulting from grating beating, might occur [56]. Therefore, a second diffuser sheet which functions also as a protective layer for the BEF films might be employed, which is usually the case in medium- and large-panel LCDs. However, for a small-panel display, because such an optical effect might not be severe and minimizing the thickness of the device is of primary concern, this second diffuser sheet might be omitted.

To further improve the axial light intensity, a reflective polarizer might be placed between the rear linear polarizer and the prism sheet. Presently, the dominating reflective polarizer is 3M´s DBEF film, which is formed by hundreds of dielectric stacks with alternating birefringent and isotropic polymer layers [53]. A detailed structure can be seen in Figure 1.46, where the isotropic layer has $n_x = n_y = n$ and the birefringent layer has $n_x = n < n_y$. Therefore, at normal incidence in the z direction, the incident wave with its polarization along the x-axis always encounters identical refractive index values and can traverse the whole system with high transmission. In contrast, light polarized along the y-axis has alternating high and low refractive indices n_H (~1.88) and n_L (~1.57), resulting in multiple internal reflections and interference effects that, in turn, affect the overall reflection and transmission. In a similar manner to a multi-layered dielectric mirror, by properly adjusting layer thickness to satisfy the Bragg condition, a reflection band with high reflectivity can be obtained. The central peak wavelength is related to the product of the thickness and refractive index of each layer. Typically, bandwidth is determined by the refractive index difference between layers. A larger refractive index difference would provide a wider reflection band

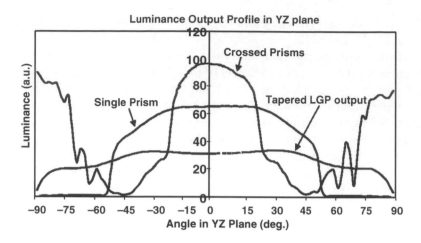

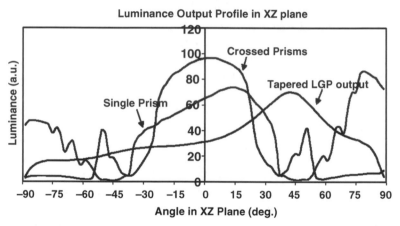

Figure 1.45 Light profile after passing the LGP and prisms (from Lighttool simulations)

and a high reflection peak within the band. To achieve a broadband DBEF film, we can laminate multiple stacks with each stack centered at a different wavelength and with a linear gradient in layer thickness [57]. With hundreds of stacks, reflectance greater than 95% can easily be obtained at normal incidence. A DBEF using multiple stacked units with a birefringent layer and an isotropic layer is quite different in its off-axis performance from a multi-layered dielectric mirror using only isotropic materials with alternating

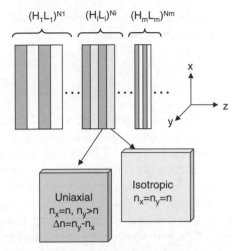

Figure 1.46 Schematic configuration of DBEF film using multiple polymeric layers

low and high refractive indices. In the multi-layered dielectric mirror, the reflection falls quickly as the incident angle departs from the normal direction, which is the Brewster effect. In the DBEF, by properly setting the n_y and n_z value for each layer, the Brewster angle for the polarization along the y-axis can be tuned to be imaginary, thus a high off-axis reflection can also be achieved [53].

Another type of reflective polarizer of interest is the nano wire grid polarizer (WGP) as illustrated in Figure 1.37 and discussed earlier in this chapter. Presently, this technology is still under development for mass production, because very narrow grid width and spacing (less than 100 nm) need to be controlled. Currently, the major fabrication methods for nano WGPs are conventional lithography, conventional nano-imprint [58, 59], roll-to-roll nano-imprint lithography (R2RNIL), and embossing [60], among which the R2RNIL method shows great potential for reducing the cost of nano WGPs.

For mobile displays, achieving minimum thickness is one of the primary requirements. To reduce the backlight unit thickness, one trend is to combine multiple film technologies into a single optical element, such as combining the prism structures into the LGP, making the bulk of the LGP a diffuser, or creating prism structures on the DBEF films [61]. Many new configurations have been developed or are under development, providing many opportunities in research, development, and business.

1.7 Summary

In this chapter, the background of transflective LCD technology has been briefly addressed. The basic device configurations and working mechanisms of the major optical elements in an LCD device, such as polarizers, LC molecular alignment, compensation films, reflectors, and backlight units, have been briefly reviewed. More detailed description of some specific topics will be provided in later chapters. This chapter not only serves as a basis for readers to understand the working principles and design concepts of a practical transflective LCD, but also paves the way for optimizing these display devices.

References

[1] Wu, S.T. and Wu, C.S. (1996) Mixed-mode twisted nematic liquid crystal cells for reflective displays. *Appl. Phys. Lett.*, **68**, 1455–1457.

[2] Wu, S.T. and Yang, D.K. (2001) *Reflective Liquid Crystal Displays*. John Wiley & Sons, Inc., New York.

[3] de Gennes, P.G. and Prost, J. (1993) *The Physics of Liquid Crystals*, 2nd edition. Clarendon Press. Oxford.

[4] Yang, D.K. and Wu, S.T. (2006) *Fundamentals of Liquid Crystal Devices*. John Wiley & Sons, Ltd, Chichester.

[5] Zhu, X., Ge, Z., Wu, T.X. and Wu, S.T. (2005) Transflective liquid crystal displays. *J. Disp. Technol.*, **1**, 15–29.

[6] For example, see BEF film from http://www.3m.com/.

[7] Takeda, A., Kataoka, S., Sasaki, T., Chida, H., Tsuda, H., Ohmuro, K., Sasabayashi, T., Koike, Y. and Okamoto, K. (1998) A super-high image quality multidomain vertical alignment LCD by new rubbing-less technology. *SID Tech. Digest*, **29**, 1077–1080.

[8] Tanaka, Y., Taniguchi, Y., Sasaki, T., Takeda, A., Koibe, Y. and Okamoto, K. (1999) A New Design to Improve Performance and Simplify the Manufacturing Process of High-Quality MVA TFT-LCD Panels. *SID Tech. Digest*, **30**, 206–209.

[9] Kataoka, S., Takeda, A., Tsuda, H., Koike, Y., Inoue, H., Fujikawa, T., Sasabayashi, T. and Okamoto, K. (2001) A New MVA-LCD with Jagged Shaped Pixel Electrodes. *SID Tech. Digest*, **32**, 1066–1069.

[10] Schadt, M. and Helfrich, W. (1971) Voltage-dependent optical activity of a twisted nematic liquid crystal. *Appl. Phys. Lett.*, **18**, 127.

[11] Soref, R.A. (1974) Field effects in nematic liquid crystals obtained with inter-digital electrodes. *J. Appl. Phys.*, **45**, 5466.

[12] Ohe, M. and Kondo, K. (1995) Electro-optical characteristics and switching behavior of the in-plane switching mode. *Appl. Phys. Lett.*, **67**, 3895.

[13] Lee, S.H., Lee, S.L. and Kim, H.Y. (1998) Electro-optic characteristics and switching principle of a nematic liquid crystal cell controlled by fringe-field switching. *Appl. Phys. Lett.*, **73**, 2881.

[14] Bos, P.J. and Koehler-Beran, K.R. (1984) The pi-cell: a fast liquid-crystal optical-switching device. *Mol. Cryst. Liq. Cryst.*, **113**, 329.

[15] Yamaguchi, Y., Miyashita, T. and Uchida, T. (1993) Wide-viewing-angle display mode for the active-matrix LCD using bend-alignment liquid-crystal cell. *SID Tech. Digest*, **24**, 273–276.

[16] Kim, K.H., Lee, K.H., Park, S.B., Song, J.K., Kim, S.N. and Souk, J.H. (1998) Domain divided vertical alignment mode with optimized fringe field effect. In Proc. 18th Int. Display Research Conf. (Asia Display ´98), pp. 383–386.

[17] Fujimura, Y., Kamijo, T. and Yoshimi, H. (2003) Improvement of optical films for high performance LCDs. *Proceedings of SPIE*, **5003**, 96–105.

[18] Yang, Y.-C. and Yang, D.K. (2008) Achromatic reduction of off-axis light leakage in LCDs by self-compensated phase retardation (SPR) film. *SID Tech. Digest*, **39**, 1955–1958.

[19] McKnight, W.H. Stotts L.B. and Monahan M.A. (1982) Transmissive and Reflective Liquid Crystal Display. U. S. Patent 4 315 258 (February 9, 1982).

[20] Yoshida, H., Tasaka, Y., Tanaka, Y., Sukenori, H., Koike, Y. and Okamoto, K. (2004) MVA LCD for Notebook or Mobile PCs with High Transmittance, High Contrast Ratio, and Wide Angle Viewing. *SID Tech. Digest*, **35**, 6–9.

[21] Pancharatnam, S. (1956) Achromatic combinations of birefringent plates. *Proc. Ind. Acad. Sci. A*, **41**, 130–144.

[22] Khoo, I.C. and Wu, S.T. (1993) *Optics and Nonlinear Optics of Liquid Crystals*. World Scientific, Singapore.

[23] Zhu, X., Ge, Z. and Wu, S.T. (2006) Analytical solutions for uniaxial-film-compensated wide-view liquid crystal displays. *J. Disp. Technol.*, **2**, 2–20.

[24] Bigelow, J.E. and Kashnow, R.A. (1977) Poincaré sphere analysis of liquid crystal optics. *Appl. Opt.*, **16**, 2090.

[25] Cognard, J. (1982) Alignment of nematic liquid crystals and their mixtures. *Mol. Cryst. Liq. Cryst., Suppl.*, **1**, 1.

[26] Mori, H. (2005) The wide view film for enhancing the field of view of LCDs. *J. Disp. Technol.*, **1**, 179.

[27] Takeuchi, K., Yasuda, S., Oikawa, T., Mori, H. and Mihayashi, K. (2006) Novel VW film for wide-viewing-angle TN-mode LCDs. *SID Tech. Digest*, **37**, 1531–1534.

[28] Lien, S.C. and John, R.A. (1994) Liquid Crystal Displays Having Multi-Domain Cells. U. S. Patent 5 309 264. (May 1994).

[29] Lien, A. and John, R.A. (1993) Multi-Domain Homeotropic Liquid Crystal Display for Active Matrix Application. EuroDisplay '93, p. 21.

[30] Lien, S.-C.A., Cai, C., Nunes, R.W., John, R.A., Galligan, E.A., Colgan, E. and Wilson, W.S. (1998) Multi-domain homeotropic liquid crystal display based on ridge and fringe field structure. *Jpn J. Appl. Phys.*, **37**, L597.

[31] Song, J.-K., Lee, K.-E., Chang, H.-S., Hong, S.-M., Jun, M.-B., Park, B.-Y., Seomun, S.-S., Kim, K.-H. and Kim, S.-S. (2004) DCCII: Novel method for fast response time in PVA mode. *SID Tech. Digest*, **35**, 1344–1347.

[32] Hanaoka, K., Nakanishi, Y., Inoue, Y., Tanuma, S. and Koike, Y. (2004) A New MVA-LCD by Polymer Sustained Alignment Technology. *SID Tech. Digest*, **35**, 1200–1203.

[33] Kim, S.G., Kim, S.M., Kim, Y.S., Lee, H.K., Lee, S.H., Lee, G.-D., Lyu, J.-J. and Kim, K.H. (2007) Stabilization of the liquid crystal director in the patterned vertical alignment mode through formation of pretilt angle by reactive mesogen. *Appl. Phys. Lett.*, **90**, 261910–261912.

[34] Matsumoto, S., Kawamoto, M. and Mizunoya, K. (1976) Field-induced deformation of hybrid-aligned nematic liquid crystals: new multicolor liquid crystal display. *J. Appl. Phys.*, **47**, 3842.

[35] Hosaki, K., Uesaka, T., Nishimura, S. and Mazaki, H. (2006) Comparison of viewing angle performance of TN-LCD and ECB-LCD using hybrid-aligned nematic compensators. *SID Tech. Digest*, **37**, 721–724.

[36] Uesaka, T., Ikeda, S., Nishimura, S. and Mazaki, H. (2007) Viewing-angle compensation of TN- and ECB-LCD modes by using a rod-like liquid crystalline polymer film. *SID Tech. Digest*, **38**, 1555–1558.

[37] Yeung, F.S., Ho, J.Y., Li, Y.W., Xie, F.C., Tsui, O.K., Sheng, P. and Kwok, H.S. (2006) Variable liquid crystal pretilt angles by nanostructured surfaces. *Appl. Phys. Lett.*, **88**, 051910.

[38] Ito, Y., Matsubara, R., Nakamura, R., Nagai, M., Nakamura, S., Mori, H. and Mihayashi, K. (2005) OCB-WV film for fast-response-time and wide-viewing-angle LCD-TVs. *SID Tech. Digest*, **36**, 986–989.

[39] Lien, A. (1990) Extended Jones matrix representation for the twisted nematic liquid-crystal display at oblique incidence. *Appl. Phys. Lett.*, **57**, 2767–2769.

[40] Yeh, P. and Gu, C. (1999) *Optics of Liquid Crystal Displays*. John Wiley & Sons, Inc., New York.

[41] Ge, Z., Wu, T.X., Zhu, X. and Wu, S.T. (2005) Reflective liquid crystal displays with asymmetric incidence and exit angles. *J. Opt. Soc. Am. A.*, **22**, 966–977.

[42] Fujimura, Y., Nagatsuka, T., Yoshimi, H., Umemoto, S. and Shimomura, T. (1992) Optical properties of retardation film. *SID Tech. Digest*, **23**, 397–400.

[43] Chen, J., Chang, K.C., Delpico, J., Seiberle, H. and Schadt, M. (1999) Wide Viewing Angle Photoaligned Plastic Films For TNLCD. *SID Tech. Digest*, **30**, 98–101.

[44] Nishimura, S. and Mazaki, H. (2003) Viewing angle compensation of various LCD modes by using a liquid crystalline polymer film 'Nisseki LC film'. *Proceedings of SPIE*, **6332**, 633203–1.

[45] Palto, S., Kasianova, I., Kharatiyan, E., Kuzmin, V., Lazarev, A. and Lazarev, P. (2007) Thin coatable birefringent films and their application to VA and IPS mode LCDs. *SID Tech. Digest*, **38**, 1563–1566.

[46] http://en.wikipedia.org/wiki/Daylight.

[47] Ting, D.-L., Chang, W.-C., Liu, C.-Y., Shiu, J.-W., Wen, C.-J., Chao, C.-H., Chuang, L.-S. and Chang, C.-C. (1999) A High Brightness and High Contrast Reflective LCD with Micro Slant Reflector (MSR). *SID Tech. Digest*, **30**, 954–957.

[48] Wen, C.-J., Ting, D.-L., Chen, C.-Y., Chuang L.-S. and Chang C.-C. (2000) Optical Properties of Reflective LCD with Diffusive Micro Slant Reflector (DMSR). *SID Tech. Digest*, **31**, 526–529.

[49] Kim, J.H., Yu, J.-H., Cheong, B.-H., Choi, Y.-S. and Choi, H.-Y. (2008) Highly efficient diffractive reflector using microgratings for reflective display. *Appl. Phys. Lett.*, **93**, 041915.

[50] Perkins, R.T., Hansen, D.P., Gardner, E.W., Thorne, J.M. and Robbins, A.A. (2000) Broadband wire grid polarizer for the visible spectrum. U. S. Patent 6 122 103 (September 2000).

[51] Yu, X.J. and Kwok, H.S. (2003) Optical wire-grid polarizers at oblique angles of incidence. *J. Appl. Phys.*, **93**, 4407–4412.

[52] Ge, Z. and Wu, S.T. (2008) Nano-wire grid polarizer for energy efficient and wide-view liquid crystal displays. *Appl. Phys. Lett.*, **93**, 121104.

[53] Weber, M.F., Stover, C.A., Gilbert, L.R., Nevitt, T.J. and Ouderkirk, A.J. (2000) Giant Birefringent Optics in Multilayer Polymer Mirrors. *Science*, **287** (5462), 2451–2456.

[54] Kim, S.S., Berkeley, B.H. and Kim, T. (2006) Advancements for highest-performance LCD-TV. *SID Tech. Digest*, **37**, 1938–1941.

[55] Scott, K. (2004) *From concept to reality to the future*. Educational presentation for the IESNA Great Lakes Region, June,

[56] Graf, J., Olczak, G., Yamada, M., Colyle, D. and Yeung, S. (2007) Backlight film & sheet technologies for LCDs. SID seminar M-12. May, 21.

[57] Li, Y., Wu, T.X. and Wu, S.T. (2009) Design optimization of reflective polarizers for LCD backlight recycling. *J. Disp. Technol.*, **5**, 335–340.

[58] Wang, J., Walters, F., Liu, X., Sciortino, P. and Deng, X. (2007) High-performance, large area, deep ultraviolet to infrared polarizers based on 40 nm line/78 nm space nanowire grids. *Appl. Phys. Lett.*, **90**, 061104.

[59] Wang, J., Chen, L., Liu, X., Sciortino, P., Liu, F., Walters, F. and Deng, X. (2006) 30-nm-wide aluminum nanowire grid for ultrahigh contrast and

transmittance polarizers made by UV-nanoimprint lithography. *Appl. Phys. Lett.*, **89**, 141105.

[60] Ahn, S.H., Kim, J.-S. and Guo, L.J. (2007) Bilayer metal wire-grid polarizer fabricated by roll-to-roll nanoimprint lithography on flexible plastic substrate. *J. Vac. Sci. Technol. B.*, **25** (6), 2388.

[61] Kalantar, K. (2006) *LCD Backlighting*. SID Application Tutorial Notes, A-5, San Francisco, June.

2

Device Physics and Modeling

2.1 Overview

A reliable modeling method for the electro-optical behavior of liquid crystal displays holds the key not only to optimizing the performance of existing displays but also to developing novel ones. Generally speaking, device modeling includes two steps: evaluation of the LC director deformation under external electric fields and calculation of the optical properties thereafter. In this chapter we will first introduce numerical methods for calculating the LC director deformation and the related optical properties, and then apply these methods to study the device physics and optics of some transflective LCD devices.

For LC director deformation calculation, the equilibrium director configuration is obtained by minimizing the total free energy of the system (either in Helmholtz form with elastic energy plus electric energy for the constant charge model or in Gibbs form with elastic energy minus electric energy for the constant voltage model [1, 2]). Popular numerical methods applied in LC device modeling are the finite difference method (FDM) [1, 3], the finite element method (FEM) [4–8], or a combination of both. FDM is mathematically simple and easy to implement, but the rectangular mesh requirement limits its application for complex structures. In contrast, FEM is quite versatile in modeling arbitrary structures with an adaptive meshing technique. In addition, as a direct solver, this method can generate accurate and fast

Transflective Liquid Crystal Displays   Zhibing Ge and Shin-Tson Wu
© 2010 John Wiley & Sons, Ltd

solutions from solving large sparse matrix systems [8]. Previously, a major obstacle to introducing FEM into LC modeling was its mathematical complexity. But as computer-aided symbolic derivation techniques have rapidly advanced, automatic FEM formula derivation for LC simulations has become possible [1, 3, 8]. At present, much commercial LCD modeling software based on the FDM or FEM method, such as LC3D [1], DIMOS [9], LCD Master [10], Techwiz LCD [11], LCD-DESIGN [12], Mouse-LCD [13], and LCQuest [14] is available on the market.

With the LC director distribution, device optical properties can then be characterized by numerical methods such as the Jones matrix method [15], the extended Jones matrix method [16–22], and the 4×4 matrix method [23–27]. The 4×4 matrix method is a complete solver of optics in 1D stratified structures, taking into account all the refractions and multiple reflections between plate interfaces. If the effect of multiple reflections is neglected (when the refractive indices have a small difference between stratified layers), the simple 2×2 matrix method is adequate for most practical purposes. In this chapter we will also briefly review the related mathematical theories of these optical methods for computing the electro-optics such as voltage-dependent transmittance and reflectance in transflective LCD devices.

To demonstrate the procedures for modeling LC directors and electro-optics, we will choose some basic transflective LCD devices as examples. We will emphasize their fundamental working mechanisms and underlying physics.

2.2 Modeling of LC Directors

2.2.1 Free Energy of Liquid Crystal Devices

To model the LC director reorientation under external electric fields, we first need to know the free energies of the device. Typically, there are three major free energies [28–30]:

1. elastic energy;
2. electric energy;
3. surface anchoring energy.

Figure 2.1 illustrates the three main types of elastic deformation: splay, bend, and twist. The elastic energy densities associated with splay, bend, and twist deformation can be expressed as $\frac{1}{2}K_{11}(\nabla \cdot \mathbf{n})^2, \frac{1}{2}K_{33}|\mathbf{n} \times \nabla \times \mathbf{n}|^2$, and

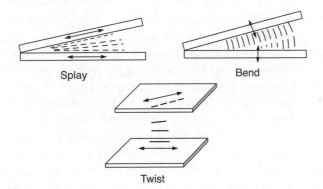

Figure 2.1 Splay, bend, and twist deformation of liquid crystals (redrawn from (28))

$\frac{1}{2}K_{22}(\mathbf{n} \cdot \nabla \times \mathbf{n})^2$, respectively. Here K_{ii} is the elastic constant and $\mathbf{n}$ represents the LC director. If the system has an inherent chirality, usually induced by mixing a chiral dopant into the nematic host, the twist energy form needs to be modified to $\frac{1}{2}K_{22}(\mathbf{n} \cdot \nabla \times \mathbf{n} + q_0)^2$ by including a chiral wave number q_0 to account for the chiral energy. Chiral wave number q_0 is also equal to $2\pi/P$, where P is the spatial pitch length for LCs to have a 2π rotation. The sign of q_0 determines the handedness of the twist. If q_0 is positive, the natural twist is right-handed, and a negative q_0 indicates a left-handed twist. Overall, the elastic energy density f_k of the LCs can be expressed as [28–30]:

$$f_k = \frac{1}{2}K_{11}(\nabla \cdot \mathbf{n})^2 + \frac{1}{2}K_{22}(\mathbf{n} \cdot \nabla \times \mathbf{n} + q_0)^2 + \frac{1}{2}K_{33}|\mathbf{n} \times \nabla \times \mathbf{n}|^2, \quad (2.1a)$$

In addition to the elastic energy resulting from the LC deformation, the free energy due to the interaction between the LC molecules and the applied electric field is also very important. The electric energy density f_e in a dielectric medium can be expressed as:

$$f_e = \frac{1}{2}\mathbf{D} \cdot \mathbf{E}. \quad (2.1b)$$

By integrating the energy density in the study space, we can derive the elastic energy $F_k = \int f_k dv$ and electric energy $F_e = \int f_e dv$. However, when evaluating the whole energy of the LC device by accounting for both elastic and electric forms, special attention should be paid. For the fixed charge model, the overall system energy to be minimized is $F = F_k + F_e$; and for a fixed voltage model, the system energy for minimization is $F = F_k - F_e$. Thurston and Berreman deduced that these two forms will, under their own conditions, go to the same variational form in minimizing the free energy of the device [2]. Nevertheless, these two free energy forms might be somewhat confusing in their expression. From thermodynamics, for the constant charge model, the system only has internal F_k and F_e without other additional external energy sources. But for the constant voltage model, to maintain a constant voltage, external work from the voltage source is needed in addition to the internal energy F_k and F_e, and that contribution is in the form $2F_e$. Considering this external work by the voltage source, the Gibbs energy of the system is $F_g = F_k + F_e - 2F_e = F_k - F_e$. However, in both cases the minimization of free energy will make the LCs reorient in such a way that the axis with a larger dielectric constant ε ($\varepsilon_{//}$ for positive $\Delta\varepsilon$ LCs, and $\varepsilon_{\perp}$ for negative $\Delta\varepsilon$ LCs) tends to be aligned along the electric field direction. In this chapter we will only derive the numerical method based on the fixed voltage case using the Gibbs energy form as an illustration of LC director simulation.

In addition to elastic energy and electric energy, surface anchoring energy is another important parameter, originating from the coupling between LC molecules and the surface alignment layer. Due to the LC and alignment material properties, there is always a preferred alignment direction for LC molecules where the surface energy is minimum, and this direction is usually designated the *easy axis* (at azimuthal angle φ_0 and polar angle θ_0), as shown in Figure 2.2. When the LC molecules are perturbed by external forces like electric fields, they will tend to align in a different orientation at the surface, say in a new direction with azimuthal angle φ and polar angle θ; thus an increased surface energy F_s will be induced. However, they still tend to align in directions where $\partial F_s/\partial\theta|_{\theta=\theta_0} = 0$ and $\partial F_s/\partial\varphi|_{\varphi=\varphi_0} = 0$. For a small deviation in the angle, the surface energy F_s can be expressed by the extended

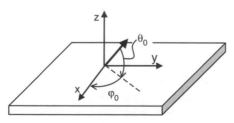

Figure 2.2 Definition of the easy axis at azimuthal angle φ_0 and polar angle θ_0

Rapini–Papoular model [31] as:

$$F_s = \frac{1}{2} W_\theta \sin^2(\theta - \theta_o) + \frac{1}{2} W_\varphi \sin^2(\varphi - \varphi_o)\cos^2(\theta_o). \qquad (2.1c)$$

From the above analysis, the overall system's Gibbs energy F_g of an LC device under a constant voltage can be expressed as:

$$F_g = \int (f_k - f_e)dv + F_s|_{rear} + F_s|_{front}. \qquad (2.2)$$

Minimizing this Gibbs energy will lead to the LC director distribution in the equilibrium state. If the surface anchoring strength is assumed to be infinity, or above a certain value called strong anchoring ($W > 10^{-3}$ J/m^2) due to mechanical rubbing, the last two surface energy terms in the simulation vanish. Other types of free energy, e.g., the energy from induced polarization of the flexoelectric effect [32, 33], might also exist and could be added into Equation (2.2).

In modeling LC directors, both vector representation, as shown in Figure 2.3, and Q tensor form [34–36] can be utilized. Detailed comparison of these two methods is provided in [35]. Owing to its mathematical simplicity, fast computing speed, and reliability (with high accuracy except in certain extreme structures with evident defects), the vector method is extensively employed in LCD simulations. Besides using n_x, n_y, and n_z for the vector $\mathbf{n}$, it can also be represented by its polar coordinates with azimuthal angle φ and polar angle θ. For 1D problems, using (φ, θ) would be very simple and straightforward. But this representation might have some oscillation

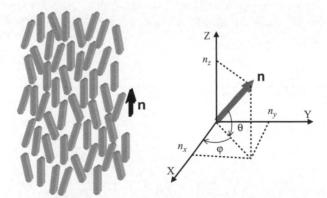

Figure 2.3 Representation of an LC director

problems when simulating LC devices under relatively high voltages, e.g., when θ approaches $90°$, φ is no longer defined. Thus, using n_x, n_y, and n_z is generally more reliable.

2.2.2 LC Simulation Flow Chart

As explained in the above section, modeling of the dynamics of LC directors in response to external voltages generally starts by minimizing the Gibbs free energy. However, from the above free energy expressions, the director and electric field profiles couple together ($f_e = \frac{1}{2}D \cdot E = \frac{1}{2}\overleftrightarrow{\varepsilon}E \cdot E$ where the dielectric tensor $\overleftrightarrow{\varepsilon}$ is a function of the LC director orientation), making it difficult to obtain a direct solution for the stable profile, except for simple 1D or 2D problems.

For a *static simulation* (targeting the final equilibrium state of LC deformation under external voltages), two minimization approaches can be adopted. The first is to use direct optimization algorithms or toolboxes like the Newton method to globally minimize the energy term in Equation (2.2) [4]. Another, more popular, approach is to utilize the Euler–Lagrange equation from the variational method [37]. The free energy F_g is a function of director values n_x, n_y, n_z and their spatial derivatives. From the Euler–Lagrange equation, minimum energy is reached when the following equations are satisfied [37]:

$$\frac{\partial f}{\partial n_l} - \frac{d}{dx}\frac{\partial f}{\partial(dn_l/dx)} - \frac{d}{dy}\frac{\partial f}{\partial(dn_l/dy)} - \frac{d}{dz}\frac{\partial f}{\partial(dn_l/dz)} = 0, \qquad (2.3a)$$

where $f = f_k - f_e$ is the free energy density and Equation (2.3a) actually has three separate equations, for $l = x$, y, and z, respectively. The unit length of the LC director should always be enforced. And the boundary conditions are:

$$\left[-\left[\frac{\partial f}{\partial(dn_l/dx)} + \frac{\partial f}{\partial(dn_l/dy)} + \frac{\partial f}{\partial(dn_l/dz)} \right] + \frac{\partial F_s}{\partial n_l} \right]_{rear} = 0, \qquad (2.3b)$$

and

$$\left[\left[\frac{\partial f}{\partial(dn_l/dx)} + \frac{\partial f}{\partial(dn_l/dy)} + \frac{\partial f}{\partial(dn_l/dz)} \right] + \frac{\partial F_s}{\partial n_l} \right]_{front} = 0. \qquad (2.3c)$$

For the boundary conditions in Equations (2.3b) and (2.3c), it is mathematically easier to revert to the (φ, θ) expression for the director, even though the bulk update still uses n_x, n_y, and n_z.

The *static* solution of the above equations is still quite complicated, especially for 2D or 3D structures. But these equations for 1D simulation can be analytically derived. The free energies can be rewritten as:

$$
f_k = \frac{1}{2}K_{11}(\nabla \cdot \mathbf{n})^2 + \frac{1}{2}K_{22}(\mathbf{n} \cdot \nabla \times \mathbf{n} + q_0)^2 + \frac{1}{2}K_{33}(|\mathbf{n} \times \nabla \times \mathbf{n}|)^2
$$

$$
= \frac{1}{2}\left(K_{11}\cos^2\theta + K_{33}\sin^2\theta\right)\left(\frac{d\theta}{dz}\right)^2 + \frac{1}{2}\left(K_{22}\cos^2\theta + K_{33}\sin^2\theta\right)\cos^2\theta\left(\frac{d\phi}{dz}\right)^2
$$

$$
- K_{22}q_0 \cos^2\theta \frac{d\phi}{dz} + \frac{1}{2}K_{22}q_0^2, \tag{2.4a}
$$

and

$$
f_e = \frac{1}{2}\varepsilon_0(\varepsilon_\perp + \Delta\varepsilon \sin^2\theta)\left(\frac{dV}{dz}\right)^2. \tag{2.4b}
$$

Substituting these bulk free energy expressions along with the surface anchoring energy from Equation (2.1c) into Equations (2.3a)(2.3c) (Euler–Lagrange equations in the form of variables φ, θ, and their spatial derivatives), the updated 1D equations can be analytically obtained. For the polar angle θ in the bulk, Equation (2.3a) becomes:

$$
-(K_{11}\cos^2\theta + K_{33}\sin^2\theta)\frac{d^2\theta}{dz^2} + (K_{33}-K_{11})\cos\theta \sin\theta\left(\frac{d\theta}{dz}\right)^2
$$

$$
+ (-2K_{22}\cos^2\theta + K_{33}\cos 2\theta)\cos\theta \sin\theta\left(\frac{d\phi}{dz}\right)^2 \tag{2.5a}
$$

$$
+ 2K_{22}q_0 \cos\theta \sin\theta \frac{d\phi}{dz} - \varepsilon_0\Delta\varepsilon \cos\theta \sin\theta\left(\frac{dV}{dz}\right)^2 = 0,
$$

and for the φ angle in the bulk, Equation (2.3a) becomes:

$$-(K_{22}\cos^2\theta + K_{33}\sin^2\theta)\cos^2\theta\frac{d^2\phi}{dz^2}$$

$$-2[(-2K_{22}\cos^2\theta + K_{33}\cos2\theta)\cos\theta\sin\theta\frac{d\theta}{dz}\frac{d\phi}{dz}\qquad(2.5b)$$

$$-2K_{22}q_0\cos\theta\sin\theta\frac{d\theta}{dz}=0.$$

Boundary conditions for the polar angle θ are:

$$\left[-(K_{11}\cos^2\theta + K_{33}\sin^2\theta)\frac{d\theta}{dz} + W_\theta\cos(\theta-\theta_0)\sin(\theta-\theta_0)\right]_{rear}=0,\qquad(2.6a)$$

and

$$\left[(K_{11}\cos^2\theta + K_{33}\sin^2\theta)\frac{d\theta}{dz} + W_\theta\cos(\theta-\theta_0)\sin(\theta-\theta_0)\right]_{front}=0.\qquad(2.6b)$$

Similarly, boundary conditions for the φ angle are:

$$\left(\begin{array}{c}-\left[(K_{22}\cos^2\theta + K_{33}\sin^2\theta)\cos^2\theta\frac{d\phi}{dz} - K_{22}q_0\cos^2\theta\right]\\ + W_\varphi\cos(\varphi-\varphi_0)\sin(\varphi-\varphi_0)\cos^2(\theta_0)\end{array}\right)_{rear}=0,\qquad(2.7a)$$

and

$$\left(\begin{array}{c}\left[(K_{22}\cos^2\theta + K_{33}\sin^2\theta)\cos^2\theta\frac{d\phi}{dz} - K_{22}q_0\cos^2\theta\right]\\ + W_\varphi\cos(\varphi-\varphi_0)\sin(\varphi-\varphi_0)\cos^2(\theta_0)\end{array}\right)_{front}=0.\qquad(2.7b)$$

Using these equations, the director profile with anchoring energy under external voltages can be computed. Nevertheless, for 2D and 3D simulations, using variables φ and θ might not be a good choice, as there could be oscillation problems. In such situations it is better to use n_x, n_y, and n_z for the bulk equation derivation, as in Equation (2.3a), and φ and θ for the boundary update.

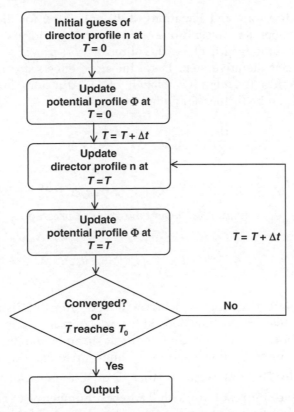

Figure 2.4 Iteration flow chart for dynamic modeling

To conduct *dynamic simulation* (targeting the LC deformation with respect to time under external voltages), an iterative process needs to be employed: minimizing the Gibbs free energy to update the LC director profile and solving the Gauss equation (equivalent to minimizing the electric energy) to update the potential profile interactively. A detailed iterative flow chart is shown in Figure 2.4. The initial LC director profile can usually be set by its boundary conditions. The stability of this iterative updating scheme is an important issue, which relies on the proper selection of the updating time step Δt [38].

It is known that minimizing the above free energy by solving the Euler–Lagrange equation and including the Rayleigh dissipation in the dynamics results in [1, 37]:

$$\frac{\partial f}{\partial n_l} - \frac{\mathrm{d}}{\mathrm{d}x}\frac{\partial f}{\partial (\mathrm{d}n_l/\mathrm{d}x)} - \frac{\mathrm{d}}{\mathrm{d}y}\frac{\partial f}{\partial (\mathrm{d}n_l/\mathrm{d}y)} - \frac{\mathrm{d}}{\mathrm{d}z}\frac{\partial f}{\partial (\mathrm{d}n_l/\mathrm{d}z)} = -\gamma\frac{\mathrm{d}n_l}{\mathrm{d}t} \qquad (2.8)$$

for the bulk directors, and Equations (2.3b) and (2.3c) for the LCs on the boundaries, where n_l represents one of n_x, n_y, and n_z, and γ is the rotational viscosity. The unit length of the LC director $n_x^2 + n_y^2 + n_z^2 = 1$ should be enforced in each iterative step. Here, the flow effects are ignored. This equation specifies the iterative relation of LC directors under external voltages, and can be further simplified to:

$$\frac{\mathrm{d}n_l}{\mathrm{d}t} + \frac{1}{\gamma}[f]_{n_l} = 0 (l = x, y, z), \tag{2.9}$$

where $[f]_{n_l} = \left(\frac{\partial f}{\partial n_l} - \frac{\mathrm{d}}{\mathrm{d}x} \frac{\partial f}{\partial(\mathrm{d}n_l/\mathrm{d}x)} - \frac{\mathrm{d}}{\mathrm{d}y} \frac{\partial f}{\partial(\mathrm{d}n_l/\mathrm{d}y)} - \frac{\mathrm{d}}{\mathrm{d}z} \frac{\partial f}{\partial(\mathrm{d}n_l/\mathrm{d}z)} \right)$. By taking a time difference $\frac{\mathrm{d}n_l}{\mathrm{d}t} = \frac{n_l^{t+\Delta t} - n_l^t}{\Delta t}$, Equation (2.9) can be further expressed as:

$$n_l^{t+\Delta t} = n_l^t - \frac{\Delta t}{\gamma}[f]_{n_l}(l = x, y, z). \tag{2.10}$$

Since the dielectric tensor $\overleftrightarrow{\varepsilon}$ includes the LC director distribution, the potential and the director functions in $[f]_{n_l}$ are coupled. Computer-assisted derivations show $[f]_{n_l}$ has a highly nonlinear form containing the highest spatial derivative of n_l with respect to x, y, and z up to the second order (e.g., $\frac{\partial^2 n_l}{\partial x \partial y}$). In order to use the first order 2D triangle or 3D tetrahedral basis function (which have the simplest form with the lowest number of variables for fast computing) to evaluate these second order derivatives, the weak form method, as discussed by Ge et al. in [8], can be introduced.

With any given LC director distribution and fixed voltage, the potential profile Φ can then be determined by solving the Gauss equation:

$$\nabla \cdot \mathbf{D} = 0 \tag{2.11}$$

The displacement $\mathbf{D}$ can be expressed by the potential Φ from $\mathbf{D} = \varepsilon_0 \overleftrightarrow{\varepsilon} \mathbf{E}$ and $\mathbf{E} = -\nabla\Phi$. Here, solving the above equation is equivalent to minimizing the electric energy $F_e = \int f_e \mathrm{d}v$ in the system from the variational method. The system region for the potential calculation defined here includes both the LC cell and other regions, such as the substrates. Either a direct solver (e.g., FEM with Ritz's method or Galerkin's method) or an iterative approach (e.g., FDM with iterations) can be applied to solve for the desired potential profile. The boundary anchoring energy can also be included. Therefore, iterations for both potential and LC director profiles can be conducted in the above

LC simulation flow chart to obtain the dynamics of LCs in response to external voltages.

2.3 Modeling of LC Optics

Once we know the LC director distribution, we can further explore the optical and electro-optical properties of LC devices. Basically, for display applications, the LC cell, along with two linear polarizers, functions like an optical valve to modulate light amplitude for different gray levels. In addition, compensation film is usually included in the display device to widen the viewing angle. Therefore the change in polarization of the light throughout the LC cell and compensation films is of primary interest. In this section the 4×4 matrix method and the extended Jones matrix method will be introduced as the main methods of characterizing the optics of display devices.

2.3.1 4×4 Matrix Method

The equivalent optical configurations of a transmissive LCD and a reflective LCD are shown in Figure 2.5(a) and 2.5(b), respectively. In each plot, the system is divided into many thin stacks, each of which can be considered a homogeneous layer. Here, we first focus on the derivation of matrix-based methods for a transmissive LCD. The simulation method for a reflective LCD can be modified from the methodology for a transmissive one by including certain special treatment [21]. In Figure 2.5(a), we consider unpolarized light entering an LCD system at an oblique angle θ_{inc}. Without losing generality, we choose a coordinate system in which the wave vector k lies in the $x - z$ plane and the incident angle in the air is designated θ_k. Here the $+z$ axis points from the rear glass substrate to the exit polarizer, and the whole LCD system is divided into N layers along the z direction. The retardation film shown in the figure can be comprised of more than one layer as needed. In some LCD devices, anti-reflection (AR) coatings are deposited on to the outer surface(s) of the polarizer(s). Under such circumstances, Layer 1 or N in Figure 2.5 represents the AR coatings instead of the polarizer.

The derivation procedures start by solving Maxwell's equations. For simplicity, we normalize the magnetic field **H** to give **E** and the new $\widehat{\mathbf{H}}$ a similar magnitude:

$$\widehat{\mathbf{H}} = \left(\frac{\mu_0}{\varepsilon_0} \right)^{1/2} \mathbf{H}. \tag{2.12}$$

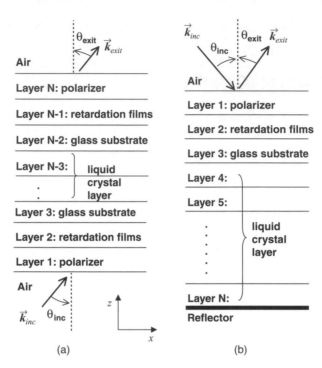

Figure 2.5 Schematic diagram of equivalent stratified structure of (a) a transmissive LCD and (b) a reflective LCD

Maxwell's equations can be expressed for $\mathbf{E}$ and $\widehat{\mathbf{H}}$ in the following forms:

$$\nabla \times \mathbf{E} = ik_0\widehat{\mathbf{H}} \tag{2.13a}$$

$$\nabla \times \widehat{\mathbf{H}} = -ik_0\overleftrightarrow{\varepsilon}\,\mathbf{E}, \tag{2.13b}$$

where $\overleftrightarrow{\varepsilon}$ is the dielectric tensor of the medium that is determined by the refractive indices and optical axis orientation of the medium (see Appendix 2.A). With $\partial/\partial y = 0$ and $\partial/\partial x = ik_x$, we can expand Equations (2.13a) and (2.13b) into six equations:

$$-\frac{\partial E_y}{\partial z} = ik_0\hat{H}_x, \tag{2.14a}$$

$$-ik_xE_z + \frac{\partial E_x}{\partial z} = ik_0\hat{H}_y, \qquad (2.14b)$$

$$ik_xE_y = ik_0\hat{H}_z, \qquad (2.14c)$$

$$-\frac{\partial \hat{H}_y}{\partial z} = -ik_0(\varepsilon_{xx}E_x + \varepsilon_{xy}E_y + \varepsilon_{xz}E_z), \qquad (2.14d)$$

$$-ik_x\hat{H}_z + \frac{\partial \hat{H}_x}{\partial z} = -ik_0(\varepsilon_{yx}E_x + \varepsilon_{yy}E_y + \varepsilon_{yz}E_z), \qquad (2.14e)$$

$$ik_x\hat{H}_y = -ik_0(\varepsilon_{zx}E_x + \varepsilon_{zy}E_y + \varepsilon_{zz}E_z), \qquad (2.14f)$$

After eliminating the longitudinal components, these six equations can be written in a matrix representation as:

$$\frac{\partial}{\partial z}\begin{pmatrix} E_x \\ E_y \\ \hat{H}_x \\ \hat{H}_y \end{pmatrix} = ik_0\, \mathbf{Q} \begin{pmatrix} E_x \\ E_y \\ \hat{H}_x \\ \hat{H}_y \end{pmatrix}, \qquad (2.15)$$

where

$$\mathbf{Q} = \begin{bmatrix} -\dfrac{\varepsilon_{zx}}{\varepsilon_{zz}}\sin\theta_k & -\dfrac{\varepsilon_{zy}}{\varepsilon_{zz}}\sin\theta_k & 0 & 1-\dfrac{\sin^2\theta_k}{\varepsilon_{zz}} \\[2ex] 0 & 0 & -1 & 0 \\[2ex] -\varepsilon_{yx}+\varepsilon_{yz}\dfrac{\varepsilon_{zx}}{\varepsilon_{zz}} & -\varepsilon_{yy}+\varepsilon_{yz}\dfrac{\varepsilon_{zy}}{\varepsilon_{zz}}+\sin^2\theta_k & 0 & \dfrac{\varepsilon_{yz}}{\varepsilon_{zz}}\sin\theta_k \\[2ex] \varepsilon_{xx}-\varepsilon_{xz}\dfrac{\varepsilon_{zx}}{\varepsilon_{zz}} & \varepsilon_{xy}-\varepsilon_{xz}\dfrac{\varepsilon_{zy}}{\varepsilon_{zz}} & 0 & -\dfrac{\varepsilon_{xz}}{\varepsilon_{zz}}\sin\theta_k \end{bmatrix}. \qquad (2.16)$$

From linear algebra, diagonalizing matrix $\mathbf{Q}$ to obtain its eigenvalues and eigenvectors would solve these coupled equations. This eigensystem can be solved by many commercial programs, such as MATLAB. In this method, we can express the diagonalized matrix $\mathbf{Q}$ as:

$$
\mathbf{Q} = \mathbf{T}
\begin{bmatrix}
q_1 & & & \\
& q_2 & & \\
& & q_3 & \\
& & & q_4
\end{bmatrix}
\mathbf{T}^{-1},
\tag{2.17}
$$

where q_1 to q_4 are mathematically the eigenvalues of $\mathbf{Q}$, and $\mathbf{T}$ consists of the corresponding eigenvectors. For simplicity, we intentionally adjust the eigenvalues q_1 to q_4 and eigenvector matrix $\mathbf{T}$ in such a way that q_1 and q_2 are positive while q_3 and q_4 are negative. For a uniaxial medium, the expression of each q can be analytically derived as:

$$
q_1 = -q_3 = (n_o^2 - \sin^2\theta_k)^{1/2},
\tag{2.18}
$$

$$
q_2 = -q_4 = -\frac{\varepsilon_{xz}}{\varepsilon_{zz}}\sin\theta_k + \frac{n_o n_e}{\varepsilon_{zz}}\left[\varepsilon_{zz} - \left(1 - \frac{n_e^2 - n_o^2}{n_e^2}\cos^2\theta\sin^2\phi\right)\sin^2\theta_k\right]^{1/2},
\tag{2.19}
$$

where θ and ϕ are the polar and azimuthal angles of a uniaxial layer, similar to those defined in Figure 2.3. Optically, q_1 and q_3 are designated the forward and backward ordinary waves, and q_2 and q_4 are designated the forward and backward extraordinary waves. With the diagonalized $\mathbf{Q}$ matrix, we can conduct a further variable transformation of the tangential field components as:

$$
\begin{bmatrix}
E_x \\
E_y \\
\hat{H}_x \\
\hat{H}_y
\end{bmatrix}
= \mathbf{T}
\begin{bmatrix}
U_1 \\
U_2 \\
U_3 \\
U_4
\end{bmatrix}.
\tag{2.20}
$$

Substituting Equation (2.17) into Equation (2.20), we obtain:

$$
\frac{\partial}{\partial z}
\begin{bmatrix}
U_1 \\
U_2 \\
U_3 \\
U_4
\end{bmatrix}
= ik_0
\begin{bmatrix}
q_1 & & & \\
& q_2 & & \\
& & q_3 & \\
& & & q_4
\end{bmatrix}
\begin{bmatrix}
U_1 \\
U_2 \\
U_3 \\
U_4
\end{bmatrix}.
\tag{2.21}
$$

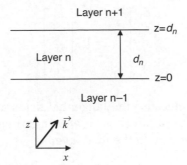

Figure 2.6 Schematic diagram of the propagation across the nth layer with a thickness of d_n in the z direction

Mathematically, Equation (2.21) is comprised of four uncoupled wave propagation equations, in which U_1 and U_2 represent the forward eigenwaves, while U_3 and U_4 represent the backward ones.

According to Figure 2.6, the solution of Equation (2.21) for each stack yields:

$$
\begin{bmatrix} U_1 \\ U_2 \\ U_3 \\ U_4 \end{bmatrix}_{n,d_n} = \mathbf{H}_n \begin{bmatrix} U_1 \\ U_2 \\ U_3 \\ U_4 \end{bmatrix}_{n,0}, \tag{2.22}
$$

where

$$
\mathbf{H}_n = \begin{bmatrix} \exp(ik_{z1}d_n) & & & \\ & \exp(ik_{z2}d_n) & & \\ & & \exp(ik_{z3}d_n) & \\ & & & \exp(ik_{z4}d_n) \end{bmatrix}, \tag{2.23}
$$

and $k_{zi} = k_0 q_i, i = 1, 2, 3, 4$. Here, we just set the rear surface to a relative coordinate $z = 0$ for simplicity, but this is not necessarily its absolute starting position when considering all the layers in the whole system. Further, going back to the relations in Equation (2.20), we can derive the relations of the tangential fields (with both forward and backward components) between the rear and front surfaces in each layer as:

$$
\begin{bmatrix} E_x \\ E_y \\ \hat{H}_x \\ \hat{H}_y \end{bmatrix}_{n,d_n} = \mathbf{P}_n \begin{bmatrix} E_x \\ E_y \\ \hat{H}_x \\ \hat{H}_y \end{bmatrix}_{n,0} , \tag{2.24}
$$

where $\mathbf{P}_n = \mathbf{T}_n \mathbf{H}_n \mathbf{T}_n^{-1}$. Because the tangential fields are continuous at each inner interface such as $z = 0^-$ and $z = 0^+$, we can correlate the tangential fields in the first layer all the way to the final layer as:

$$
\begin{bmatrix} E_x \\ E_y \\ \hat{H}_x \\ \hat{H}_y \end{bmatrix}_{N,d_N^-} = \mathbf{P}_N \mathbf{P}_{N-1} \cdots \mathbf{P}_2 \mathbf{P}_1 \begin{bmatrix} E_x \\ E_y \\ \hat{H}_x \\ \hat{H}_y \end{bmatrix}_{1,0^+} = \mathbf{P} \begin{bmatrix} E_x \\ E_y \\ \hat{H}_x \\ \hat{H}_y \end{bmatrix}_{1,0^+} . \tag{2.25}
$$

Please pay special attention here to the subscripts and superscripts for the position indices of the first and final layers, as the above continuous relation does not hold at the incident and exit surfaces due to the index mismatch there. In other words, our final target is to correlate the incident electric fields $[E_x^+, E_y^+]_{1,0^-}$ at the $z = 0^-$ (absolute coordinate) and the exit (or output) electric fields $[E_x^+, E_y^+]_{N,D^+}$ at $z = D^+$ (here D is the total thickness of the system or the final coordinate of the exit surface) as follows:

$$
\begin{bmatrix} E_{x,}^+ \\ E_y^+ \\ E_x^- \\ E_y^- \end{bmatrix}_{N,D^+} = \mathbf{M} \begin{bmatrix} E_x^+ \\ E_y^+ \\ E_x^- \\ E_y^- \end{bmatrix}_{1,0^-} . \tag{2.26}
$$

Two additional correlation matrices, $\mathbf{C}_0$ for $z = 0$ and $\mathbf{C}_{N+1}$ for $z = D$, need to be included to make $\mathbf{M} = \mathbf{C}_{N+1} \cdot \mathbf{P} \cdot \mathbf{C}_0$. From Huang's derivation [27], they have the following forms:

$$
\mathbf{C}_0 = \begin{bmatrix} 1 & 0 & 1 & 0 \\ 0 & 1 & 0 & 1 \\ 0 & -\sqrt{\varepsilon_{rz}} & 0 & \sqrt{\varepsilon_{rz}} \\ \dfrac{\varepsilon_r}{\sqrt{\varepsilon_{rz}}} & 0 & -\dfrac{\varepsilon_r}{\sqrt{\varepsilon_{rz}}} & 0 \end{bmatrix} , \tag{2.27}
$$

and

$$C_{N+1} = \begin{bmatrix} \dfrac{1}{2} & 0 & 0 & \dfrac{\sqrt{\varepsilon_{rz}}}{2\varepsilon_r} \\[2ex] 0 & \dfrac{1}{2} & -\dfrac{1}{2\sqrt{\varepsilon_{rz}}} & 0 \\[2ex] \dfrac{1}{2} & 0 & 0 & -\dfrac{\sqrt{\varepsilon_{rz}}}{2\varepsilon_r} \\[2ex] 0 & \dfrac{1}{2} & \dfrac{1}{2\sqrt{\varepsilon_{rz}}} & 0 \end{bmatrix}, \qquad (2.28)$$

where $\varepsilon_r = n^2$ is the permittivity of the material and ε_{rz} is its component in the z direction. From Equation (2.26), we can correlate the input and output fields, which can be further processed to analyze the system characteristics such as transmittance, bandwidth, and many others. Using numerical solvers to obtain the eigenvalues and eigenvectors of the characteristic 4×4 matrix **Q**, the above derivation is quite straightforward and easy to implement. In addition, the 4×4 matrix method accounts for both forward and backward waves in the simulation by constructing a complex matrix for each single stack. Hence, this method provides the exact solution of Maxwell's equations for 1D optical problems. By a superposition of all waves, which might be in a constructive or destructive way, the outputs might have small oscillations. To eliminate this small oscillation, a local averaging of outputs with respect to the wavelengths needs to be performed in calculations, leading to an increase in computing time.

2.3.2 2 × 2 Extended Jones Matrix Method

In a typical LCD device, except at the air–polarizer interface, the refractive indices of adjoining layers are quite close to each other. Thus, the backward or reflective waves are usually negligible. In other words, in calculating the wave propagation through the LCD system, accounting for only the forward or transmissive waves is adequate to characterize light polarization changes. Based on this fact, 2×2 matrix methods have been developed to compute the optics of LCDs; these output adequately accurate results in a short computing time. A lot of work has been done to derive the matrix forms for both uniaxial and even biaxial media [16–22]. In this section we will use a different way to derive the 2×2 matrix method based on the above 4×4 matrix method analysis.

Based on the assumption that the backward waves are negligible, referring to Equation (2.20), U_3 and $U_4 \sim 0$. Therefore we can rewrite the 4×4 matrix $\mathbf{T}$ as four 2×2 matrices in the following form:

$$\mathbf{T} = \begin{bmatrix} \mathbf{T}_{11} & \mathbf{T}_{12} \\ \mathbf{T}_{21} & \mathbf{T}_{22} \end{bmatrix}, \tag{2.29}$$

With U_3 and U_4 neglected, we can further obtain:

$$\begin{bmatrix} E_x \\ E_y \end{bmatrix} \approx \begin{bmatrix} E_x^+ \\ E_y^+ \end{bmatrix} = \mathbf{T}_{11} \begin{bmatrix} U_1 \\ U_2 \end{bmatrix}, \tag{2.30}$$

and

$$\begin{bmatrix} \hat{H}_x \\ \hat{H}_y \end{bmatrix} \approx \begin{bmatrix} \hat{H}_x^+ \\ \hat{H}_y^+ \end{bmatrix} = \mathbf{T}_{21} \begin{bmatrix} U_1 \\ U_2 \end{bmatrix}. \tag{2.31}$$

Solving the propagation wave equations for U_i ($i = 1$ and 2), we can get the electric fields correlated between the rear and front surfaces, as shown in Figure 2.6, as:

$$\begin{bmatrix} E_x \\ E_y \end{bmatrix}_{n,d_n} = [\mathbf{T}_{11}]_n \begin{bmatrix} \exp(ik_{z1}d_n) & \\ & \exp(ik_{z2}d_n) \end{bmatrix} [\mathbf{T}_{11}^{-1}]_n \begin{bmatrix} E_x \\ E_y \end{bmatrix}_{n,0} = \mathbf{J}_n \begin{bmatrix} E_x \\ E_y \end{bmatrix}_{n,0}. \tag{2.32}$$

And for all the layers in the stacked optical structure:

$$\begin{bmatrix} E_x \\ E_y \end{bmatrix}_{N,D^-} = \mathbf{J}_N \mathbf{J}_{N-1} \cdots \mathbf{J}_2 \mathbf{J}_1 \begin{bmatrix} E_x \\ E_y \end{bmatrix}_{1,0^+}. \tag{2.33}$$

The index mismatch at the entering and exit interfaces is quite evident. Thus, to correlate the incident electric fields at $z = 0^-$ (absolute coordinate) and the exit (or output) electric fields at $z = D^+$, additional matrices are required. In other words, we need to correlate the incident wave components $[E_x^+, E_y^+]^{\mathrm{T}}$

from air to the display (at $z = 0^-$) and the exit ones $[E_x^+, E_y^+]^T$ from the display to air (at $z = D^+$) in the form:

$$
\begin{bmatrix} E_x^+ \\ E_y^+ \end{bmatrix}_{N,D^+} = J_{Ext} J_N J_{N-1} \cdots J_2 J_1 J_{Ent} \begin{bmatrix} E_x^+ \\ E_y^+ \end{bmatrix}_{1,0^-}, \tag{2.34}
$$

where

$$
J_{Ent} = \begin{pmatrix} \dfrac{2\cos\theta_p}{\cos\theta_P + n_P\cos\theta_k} & 0 \\ 0 & \dfrac{2\cos\theta_k}{\cos\theta_k + n_P\cos\theta_P} \end{pmatrix}, \tag{2.35}
$$

$$
J_{Ext} = \begin{pmatrix} \dfrac{2n_p\cos\theta_k}{\cos\theta_P + n_P\cos\theta_k} & 0 \\ 0 & \dfrac{2n_p\cos\theta_p}{\cos\theta_k + n_P\cos\theta_P} \end{pmatrix}. \tag{2.36}
$$

Here, n_p is the real part of the averaged refractive indices of the outermost polarizers and $\theta_P = \sin^{-1}(\sin(\theta_k)/n_p)$. If the air–display interface layer is not a polarizer, then n_p should be adjusted to the new refractive index of the material. The relation between incident and exit beams can be further used to characterize the transmittance of the display [22].

In the above derivation of a 2×2 matrix method, we adopt the results from 4×4 matrix method analysis. However, analytical expressions of extended Jones matrix methods have been well developed [20]. For reference, we also list the forms of these equations in Appendix 2.A, and readers can utilize the forms from either method. However, one limitation of the analytical solutions of the extended Jones matrix method [20] is that they are restricted to uniaxial media. The method derived above can be applied to both uniaxial and biaxial media, when the dielectric tensor $\overset{\leftrightarrow}{\varepsilon}$ is updated to the biaxial form.

2.3.3 Numerical Examples

Using the derivations in both 4×4 and 2×2 matrix-based methods, several numerical examples will be provided in this section. The first example involves calculating the Bragg reflection spectrum of a cholesteric liquid crystal (CLC) cell using the 4×4 matrix method. The second example computes the voltage-dependent curves of a reflective MTN cell under circular polarizers. Additional electro-optics can be characterized by the extension of these methods.

CLC is an LC phase with chiral molecules or mixtures containing chiral dopant. In a CLC cell, there is a helical structure where neighboring directors are twisted slightly with respect to each other. The distance along the helical axis for the director to have a 2π rotation is called the pitch P_0. For a CLC cell in a planar texture with a given pitch P_0, there is a periodic structure of the reflective index in the cell at normal direction and the period is $P_0/2$ due to the symmetry between director $\mathbf{n}$ and $-\mathbf{n}$. For unpolarized incident light, circular polarization with the same handedness as the CLC cell will be strongly reflected because of the constructive interference of the light reflected from different positions of the cell, while circular polarization with the opposite handedness will have a high transmission. Bragg reflection occurs at the central wavelength $\lambda_0 = <n> \cdot 2\text{Period} = <n> \cdot 2(P_0/2) = <n> \cdot P_0$, with a bandwidth given by $\Delta n \cdot P_0$, where $<n>$ and Δn are the averaged refractive index and the birefringence of the nematic-like LC materials in a CLC cell, respectively. To describe the overall reflection or transmission, all the internal reflections must be considered. As a complete solution to the Maxwell equations in 1D optics, the 4×4 matrix method is a good candidate solver. We applied the above-derived 4×4 matrix method to calculate the reflection spectrum of a CLC cell with $P_0 = 344.6$ nm (cell thickness is tuned at $d \sim 12P_0$), and $\Delta n \sim 0.227$ at $\lambda = 550$ nm and $n_o \sim 1.48$. In the calculation, an AR coating to eliminate surface reflections was numerically included. The results are shown in Figure 2.7, and exhibit good agreement with the theoretical prediction.

Our second numerical example uses the 2×2 extended Jones matrix method to study the electro-optics of a reflective LCD. To extend the above-derived methods to simulate a reflective LCD, different methods can be applied. One simple and direct method is the so-called mirror-image, or unfolding, method [22]. The reflective LCD configuration shown in Figure 2.5 can be transferred to an equivalent transmissive configuration, where mirror image stacks are unfolded behind the original ones, as shown in Figure 2.8. The mirror image stacks are functions of the incident wave, and the original stacks are functions of the reflected exit wave. Therefore, the wave vector component k_x in the front half of the original stacks is $k_{exit} \cdot \sin(\theta_{exit})$ and the optical axis orientation of each layer is just $\vec{n}_1$ at (ϕ, θ), as shown in Figure 2.8. The wave vector component k_x in the mirror image stacks results from the incident wave as $k_{inc} \cdot \sin(\theta_{inc})$. In addition, the optical axis orientation is $\vec{n}_2$ at $(\phi, -\theta)$ for a layer in the mirror image stacks whose corresponding front layer

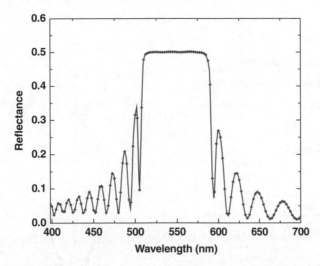

Figure 2.7 Reflection spectrum of a CLC cell centered at $\lambda = 550$ nm

is at $\vec{n}_1$. Because of the optical symmetry, $\vec{n}_2$ at $(\phi, -\theta)$ is equivalent to $\vec{n}_3$ at $(\phi + \pi, \theta)$. If the incident angle θ_{inc} is equal to θ_{exit}, the reflector can be treated as a void layer, otherwise a special correlation matrix needs to correlate the incident wave and reflected wave upon the reflector surface, such as on a slanted reflector surface [39]. In addition, considering energy flow conservation, the incident wave and exit wave might have different cross-sections. Certain correction coefficients must also be taken into consideration. For more details on simulating a reflective LCD with asymmetrical incident and exit angles, readers can refer to [22].

To illustrate the 2×2 matrix method derived above, we give a numerical example concerning reflective normally white mixed-mode twisted nematic (MTN) cells [40]. For the MTN mode simulations, we choose ZLI-4792 (from Merck) as the LC material, the parameters of which are: $n_e = 1.5763$, $n_o = 1.4794$, $\varepsilon_{//} = 8.3$, $\varepsilon_{\perp} = 3.1$, $K_{11} = 13.2$ pN, $K_{22} = 6.5$ pN, and $K_{33} = 18.3$ pN. To obtain a dark state, a circular polarizer is placed in front of the MTN cell. A commonly employed circular polarizer is shown in Figure 2.9; this consists of a linear polarizer, a monochromatic half-wave plate, and a monochromatic quarter-wave plate. The mechanism of this circular polarizer has been discussed in Chapter 1. Here, n_e for the quarter-wave film is 1.5110 and n_o is 1.5095, with a film thickness of 91.67 µm (at $\lambda = 550$ nm). On the other hand, the half-wave film has $n_e = 1.5123$ and $n_o = 1.5089$ with a film thickness of 80.88 µm (at $\lambda = 550$ nm). Here, two MTN cells are employed: a 90° MTN cell with $d\Delta n = 240$ nm and a 63.6° MTN cell with $d\Delta n = 192$ nm.

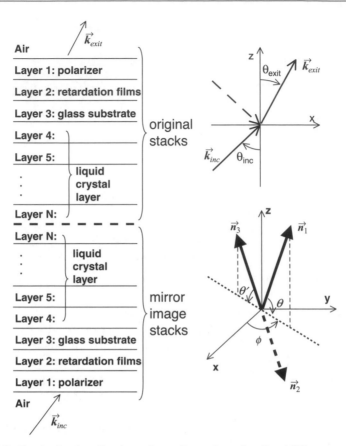

Figure 2.8 Equivalent optical configuration of a reflective LCD represented by a transmissive mode

Figure 2.10 shows the simulated VR curves of the normally white 90° and 63.6° MTN cells discussed above. In the simulations, the LC pre-tilt angle is assumed to be 2° and LC cell thicknesses are set at 2.42 μm and 1.94 μm in accordance with the above-mentioned retardation values, respectively. As we can see from Figure 2.10, the 90° MTN exhibits a good dark state at a much lower driving voltage, resulting from the self-compensation of orthogonal LC alignment at two boundaries. But its reflectance in the voltage-off state is only about 88%. For the 63.6° MTN cell, its maximum reflectance can reach almost 100% but the dark state voltage is over 8 V_{rms}, at which there is still a small amount of light leakage resulting from the overall phase retardation. To reduce light leakage and lower the driving voltage,

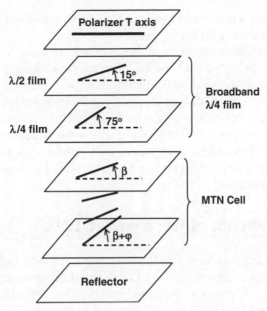

Figure 2.9 Optical configuration of an MTN cell with a broadband circular polarizer; β is the angle between the front LC rubbing direction and the polarizer transmission axis, and φ is the twist angle

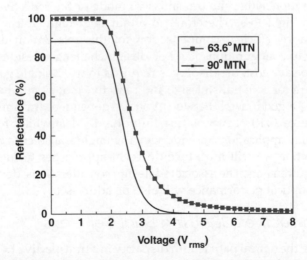

Figure 2.10 Simulated VT curves of a 63.6° MTN and a 90° MTN cell with a circular polarizer

compensation films are needed to compensate the residual phase originating from the boundary layers.

The above optical methods can be applied to both transmissive and reflective LCDs, provided that the LC director profiles have been obtained. As well as voltage-dependent transmittance or reflectance curves, these methods can also be employed to study other important electro-optical properties like light polarization changes through the LCD system, iso-contrast, and color performance. The theories and methods derived for both LC director distribution and LC optics are good preparation for studying and understanding the device physics and optics related to transflective LCDs, which will be discussed in the next section.

2.4 Device Physics of Transflective LCDs

With the numerical simulation methods in hand, we will continue to discuss the device physics of transflective LCDs. By combining the transmissive and reflective modes into a single device, a direct problem is the phase retardation mismatch between the backlight in the transmissive sub-pixel (traversing the LC cell once) and the ambient light in the reflective sub-pixel (traversing the LC cell twice). Adopting different cell gaps could solve this problem, but the manufacturing is quite difficult and its cost is high. It would be highly desirable to have a uniform cell gap. On the other hand, in a uniform cell configuration, either the transmissive mode or the reflective mode will suffer heavy light loss. Therefore, to attain both high transmittance and high reflectance, efficient compensation for this retardation difference in single-cell-gap configurations is a key design challenge. Besides good light efficiency, to save cost, the display is required to use a single gamma curve driving circuit for both transmissive and reflective modes simultaneously. In other words, a good match between voltage-dependent transmittance (VT) and reflectance (VR) curves is also important. Meanwhile, for high-end mobile display applications, a wide viewing angle is also critical.

In this section we will introduce different approaches to fulfilling these practical requirements. The associated design considerations, device physics, and electro-optical performance will also be addressed.

2.4.1 Transflective LCDs Using Dual Cell Gaps

To overcome the optical path length disparity in a transflective LCD, the most straightforward approach is to use dual cell gaps [41, 42]. The device configuration is shown in Figure 2.11. In this device, each pixel is divided

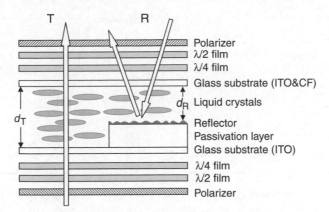

Figure 2.11 Device configuration of a dual-cell-gap transflective LCD

into two sub-pixels: a transmissive region and a reflective region, where the cell gap d_T in the transmissive region is about twice that of d_R in the reflective region. The LC alignment for this configuration can be a pure phase retardation type like a homogeneous cell, a pi-cell, or a vertical alignment cell. To obtain a dark state for the reflective sub-pixel, a circular polarizer is placed in front of the LC cell. A second circular polarizer is placed behind the LC cell to obtain a common dark state for the transmissive sub-pixel. For broadband operation, the circular polarizer consists of a linear polarizer, a half-wave plate, and a quarter-wave plate at specific orientations. To reduce cost, a conventional circular polarizer with a linear polarizer and a monochromatic quarter-wave plate is also widely used.

Referring to the LC mode, if vertical alignment is employed in the dual-cell-gap configuration, both transmissive and reflective modes exhibit a good dark state so that the contrast ratio is high. But for the normally white mode using a homogeneous cell or a pi-cell, compensation film is required. As discussed in Chapter 1, for a homogeneous cell or pi-cell, the LC molecules near the surfaces are usually strongly anchored, resulting in some residual phase retardation even at a relatively high voltage. In other words, the residual phase retardation may cause some light leakage, reducing the contrast ratio. Figure 2.12(a) plots the voltage-dependent phase retardation values of a 4.2 μm homogeneous cell for the transmissive region and a 2.1 μm cell in the reflective region. The LC material used here is MLC-6614 with $\Delta n \sim 0.0787$ at $\lambda = 550$ nm. Even at 6 V_{rms}, the phase retardation ($2\pi d\Delta n/\lambda$) of the reflective sub-pixel and the transmissive sub-pixel is still $\sim 0.06\pi$ and 0.12π, respectively. Consequently, the VT and VR curves without compensation cannot

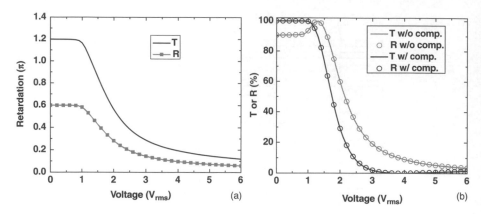

Figure 2.12 (a) Voltage-dependent phase retardation for the T and R modes, and (b) VT and VR curves without compensation films for a dual-cell-gap TR-LCD

reach a good dark state, as shown in Figure 2.12(b), although their VT and VR curves still overlap quite well.

Fortunately, the residual phase retardation of the LC cell in the voltage-on state can be compensated by the quarter-wave plates, as shown in Figure 2.11, instead of introducing additional films. In the optical configuration, to minimize the dependence on wavelength dispersion of the dark state, two linear polarizers are usually set perpendicular to each other and the two half-wave (or quarter-wave) plates in the different circular polarizers are also crossed. To compensate for the dark state of the homogeneous cell at a relatively high voltage, we need to first focus on the reflective sub-pixel, rather than the transmissive sub-pixel which has more optimization freedom by counting in the rear stacks. First, the residual retardation at a preferred operation voltage, say 4 V_{rms}, is obtained from the reflective curve in Figure 2.12(a). If the front quarter-wave plate has its optical axis parallel to (or perpendicular to) the front LC rubbing direction, this residual retardation is then deducted from (or added to) the phase retardation in the quarter-wave plate. Hence, at this desired operating voltage, the reflective LC cell and the adjusted 'quarter-wave' plate together form a quarter-wave plate for a good dark state. We then go on to optimize the transmissive sub-pixel by adjusting the phase retardation value of the rear quarter-wave plate. For example, the transmissive LC cell's residual retardation is partially compensated by the front quarter-wave plate. And the rear quarter-wave plate should be adjusted so that its phase retardation, together with the remaining transmissive residual phase retardation, functions as a new rear quarter-wave plate. Based on this process, the newly adjusted VT and VR curves are also plotted in Figure 2.12(b), where the operating voltage is selected to be 4 V_{rms}. As we can see from the new plots, the VT and VR curves also overlap with each other quite well, making a single gamma curve driving possible. We can also select a lower driving voltage such as 3.3 V_{rms} by adjusting the compensation values of the quarter-wave plates. A low driving voltage is highly desirable for mobile displays because it prolongs battery life.

The quarter-wave plates are uniaxial A-plates whose molecules are stretched to align parallel to the substrate, while the LC boundary directors have a hybrid alignment in the dark state (as discussed in Chapter 1). Although the rear and front quarter-wave plates can cancel the residual phase retardation from the LC boundary directors along the axial direction, their off-axis phase retardation change tendencies (e.g., with respect to changes in the polar angle) are distinct. As a result, their off-axis phase retardation compensation will deteriorate significantly, leading to significant light leakage. Figure 2.13 shows the iso-contrast plot at $\lambda = 550$ nm for the transmissive mode in a transflective LCD using a homogeneous cell and conventional broadband circular polarizers. This single-domain LC profile makes the viewing angle quite asymmetrical and narrow. The 10 : 1 contrast ratio is confined to less than 30° in most directions.

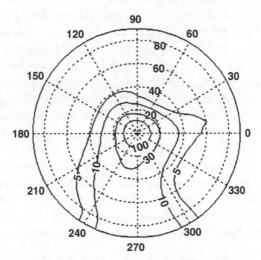

Figure 2.13 Iso-contrast plot of the transmissive ECB cell under conventional broadband circular polarizers

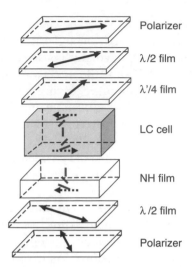

Figure 2.14 Optical configuration (T sub-pixel) of the NH film compensated homogeneous transflective LCD

To improve the viewing angle, an NH film with a hybrid alignment is used to compensate the off-axis phase retardation from the boundary LC layers in the dark state [43, 44]. The optical configuration and the related compensation mechanism are shown in Figure 2.14. Due to anti-parallel rubbing, the optical axes of the LC directors in both the rear and front cell regions under relatively high voltage show a similar inclination. An NH film with a hybrid alignment is placed behind the LC cell with its directors aligned parallel to the rear LC rubbing direction but in an opposite inclination direction. In the normal direction, the phase retardation of the NH film is similar to the adjusted quarter-wave plate, as discussed previously. But for off-axis incidence, the NH film's phase retardation change with respect to polar angle variation can be adjusted to approach that of the LC cell. Therefore, the off-axis light leakage can be significantly suppressed and the viewing angle greatly increased, as shown in Figure 2.15. But the single domain from the LC cell still makes the viewing angle a little asymmetrical. The average tilt angle of the NH film directors and the optical axis of each retardation film in the device need to be investigated to obtain optimal results. From simulation, a high average tilt angle is more helpful in increasing the viewing angle, and the angle between the slow axis of the half-wave plate and the transmission axis of the polarizer is around 30°.

In contrast to the above designs which use homogeneous alignment, vertical alignment leads to a good dark state for both transmissive and

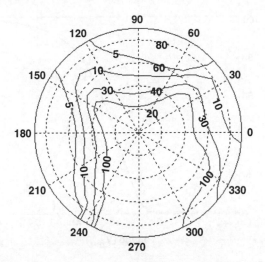

Figure 2.15 Iso-contrast plot (T sub-pixel) of the NH film compensated transflective LCD

reflective modes when no voltage is applied. Hence, its optical compensation is usually less complicated. Because the reflective mode is mainly used under strong ambient light, the major design focus of a transflective LCD is to optimize the transmissive mode, and the optical performance of the reflective mode is often compromised. For example, instead of using broadband circular polarizers, a conventional circular polarizer consisting of a linear polarizer and a single monochromatic quarter-wave plate can be used to reduce costs. By using a conventional circular polarizer, it also becomes optically easier to compensate to obtain a wide viewing angle. A detailed introduction to wide-view and broadband circular polarizer designs will be given in Chapter 3. With present fabrication technologies, it is relatively easy to form multiple domains in VA configurations such as MVA or PVA. However, there are some drawbacks associated with MVA and PVA, such as a reduced aperture ratio and severe surface pooling, as discussed in Chapter 1.

2.4.2 Transflective LCDs Using Dual Gamma Curves

A dual-cell-gap transflective LCD exhibits good light efficiency and a single gamma curve, but it requires different cell thicknesses for the transmissive and reflective regions. As a result, it has a complicated fabrication process

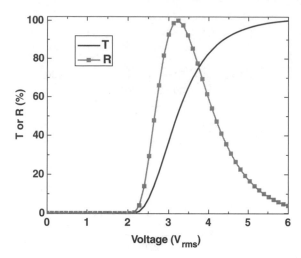

Figure 2.16 VT and VR curves from a single-cell-gap transflective VA LCD

which, in turn, increases the cost. In addition, the abrupt change in LC distribution near the boundary reduces the contrast ratio of the display. Therefore, a single-cell-gap configuration is always preferred from fabrication and performance viewpoints. Nevertheless, by using the same cell thickness, different phase retardations make the VT and VR curves diverge, as shown in Figure 2.16. If the operating voltage is optimized for the transmissive mode at $V = 6$ V_{rms}, the gray level control for the reflective mode will be wrong. On the other hand, if the operating voltage is biased at $V = 3.3$ V_{rms} for the reflective mode, the transmissive mode will experience a significant loss in light efficiency.

To solve this problem, the most direct method is to adopt separate gamma curves for the transmissive and reflective modes simultaneously [45]. The device configuration is shown in Figure 2.17, where each pixel has two TFTs to apply different voltages to the transmissive and reflective sub-pixels. Since two TFTs can be fabricated by using the same or similar photolithography steps, the related cost increase is tolerable. The major drawback is the reduced aperture ratio of the display. Although these TFTs can be buried under the metal reflector, the additional opaque gate lines and data lines still occupy a greater area than in designs using only one TFT. Therefore, in this configuration, to further increase the aperture ratio, these two TFTs need to share one gate line or one data line, and the driving scheme needs to be redesigned to accommodate the driving sequences accordingly.

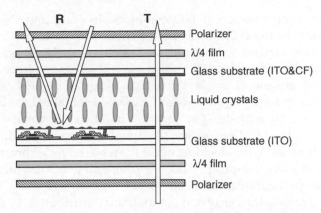

Figure 2.17 Device configuration of a single-cell-gap transflective LCD using two TFTs

2.4.3 Transflective LCDs Using Dual Electric Fields

As an alternative to the dual-cell-gap approach, it is also possible to generate different electric fields in the transmissive and reflective regions to control the LC reorientation and then balance the phase retardation [46, 47]. For example, the device shown in Figure 2.18 utilizes patterned conductive reflectors in front of a planar common electrode to create stronger electric fields in the transmissive region. In the transmissive region, the LC layer is sandwiched

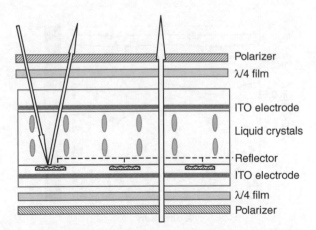

Figure 2.18 Schematic of a transflective LCD using commonly biased reflectors

between the front transparent ITO common electrode and the rear ITO pixel electrode. In the reflective region, periodically patterned conductive layers are formed in front of the rear pixel electrode as the reflectors, which are commonly biased to a certain voltage (say, $V = 0$ or a voltage below the threshold voltage) for all pixels. When no voltage is applied, all LCs are aligned vertically so that a very good dark state is obtained for both transmissive and reflective modes. On the other hand, as the voltage exceeds a threshold (V_{th}), the LC directors are tilted and the phase retardation effect takes place. Because of the shielding effect from biased reflectors, the electric field in the reflective sub-pixel is much weaker and generates less tilt in the LCs than that in the transmissive sub-pixel.

Figure 2.19 shows the calculated equipotential distribution in the LC cell region within one repetitive period, when $V = 6.5 \, V_{rms}$ is applied to the rear pixel electrode and all the reflectors are biased at $0 \, V_{rms}$. The potential difference between neighboring vertical lines is $0.25 \, V_{rms}$. Here, the LC material employed is a negative dielectric anisotropy ($\Delta\varepsilon$) Merck mixture MLC-6608 with the rubbing direction parallel to the lengthwise direction of the strip reflectors. The reflectors have a width $w = 2 \, \mu m$ and a gap $g = 5 \, \mu m$. To demonstrate a relative position, both vertical and horizontal dimensions are normalized. In the transmissive regions ($x \in [0,0.18]$, $[0.32, 0.68]$, and $[0.82, 1]$), the potential values decrease quite uniformly from the rear electrode to the front one in the vertical direction. But the starting potential from the rear LC surface is reduced to about $4.5 \, V_{rms}$, compared with $6.5 \, V_{rms}$ on the rear pixel electrode. In the reflective regions ($x \in [0.18, 0.32]$ and $[0.68, 0.82]$),

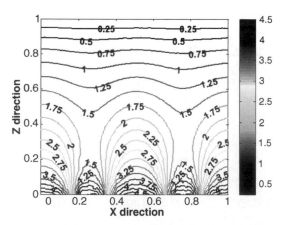

Figure 2.19 Equipotential line profile in the proposed transflective LCD when a relatively high voltage ($V = 6.5 \, V_{rms}$) is applied (redrawn from (47))

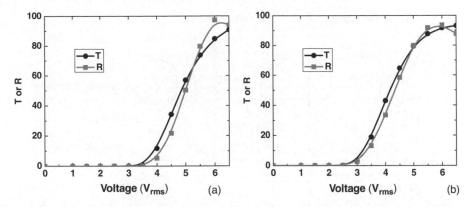

Figure 2.20 VT and VR curves at the commonly biased voltage of (a) 0 V_{rms} with $w=2\,\mu m$ and $g=5\,\mu m$, and (b) 1.5 V_{rms} with $w=3\,\mu m$ and $g=6\,\mu m$ (redrawn from (47))

uniform potential change in the vertical direction only occurs near the surfaces, e.g., from $z=0$ to ~0.2 and from $z\sim0.5$ to 1.0. A region with ultra-low field components is formed in the bulk. Under such a field distribution, the LC directors in the transmissive region can tilt more than those in the reflective region. This means that the effective phase retardation in the transmissive region is more than that in the reflective region, making it possible to match the VT and VR curves.

The simulated VT and VR curves are plotted in Figure 2.20(a) when the commonly biased voltage at the reflectors is 0 V_{rms}. At an operating voltage of 6.5 V_{rms}, both transmissive and reflective modes could reach a high normalized light efficiency of about 90% (with respect to the transmittance from two parallel linear polarizers). Drawbacks are the high threshold and high operating voltage, which mainly originate from the shielding effect of the biased reflectors. To reduce driving voltage, we could use a thinner passivation layer between the pixel electrode and the patterned reflectors. We could also increase the bias voltage on the reflectors from 0 V_{rms} to 1.5 V_{rms}. Figure 2.20(b) shows the VT and VR curves at the new bias voltage, and the driving voltage is reduced to about 2.4 V_{rms}. A better match between the VT and VR curves is also obtained. Because the bias voltage is below threshold, the dark state will not be affected.

In this example, different electric fields between the transmissive and reflective regions are obtained by shielding the voltage from a metal layer (the reflector). In yet another approach, the electric field difference can be adjusted by forming different patterned slits in the vertical alignment cell [48].

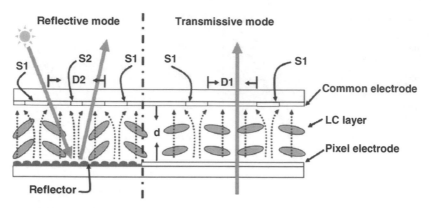

Figure 2.21 Device configuration of a transflective LCD using different slit widths in the transmissive and reflective regions (redrawn from (48))

In Figure 2.21, the transmissive region has a slit width $S_1 \sim 4\,\mu m$, and slit gap $D_1 \sim 18\,\mu m$, and the reflective region has the same slit width $S_1 \sim 4\,\mu m$ but a different slit gap $D_2 \sim 12\,\mu m$. Due to the reduced size of the slit gap in the reflective region, a fine slit $S_2 \sim 2\,\mu m$ is formed in the center of the slit gap D_2 to avoid domain breakup. On the rear substrate, both the reflector and the pixel electrode are planar in shape and have no slits. Therefore, a larger slit gap in the transmissive region generates electric fields with stronger vertical components. In contrast, the smaller gap size in the reflective region generates more fringe electric fields there. It is possible to compensate for the optical path difference between these two regions. The simulated VT and VR curves are shown in Figure 2.22, where the maximum possible light efficiency from the two chosen polarizers is about 35%. This design exhibits reasonably good light efficiency and a good match between the VT and VR curves. In addition, the slit-formed multi-domain configuration also helps to widen the viewing angle of the display. As will be discussed in Chapter 4, this configuration will have good application potential if polymer sustained surface alignment is further adopted.

Another successful method to compensate for the optical path difference between the regions is to form a voltage-shielding capacitor placed in series with the LC cell in the reflective region, while leaving the LC cell in the transmissive region directly sandwiched between the driving electrodes. This device concept was first proposed by Roosendaal *et al.* from Philips [49] and was later reported to have been implemented in a transflective VA LCD by Kang *et al.* [50]. The device configuration and its equivalent circuit are shown in Figure 2.23. In the reflective region, a passivation layer with a thickness of

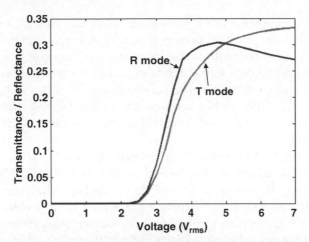

Figure 2.22 VT and VR curves of a transflective LCD using different slit widths in the transmissive and reflective regions (redrawn from (48))

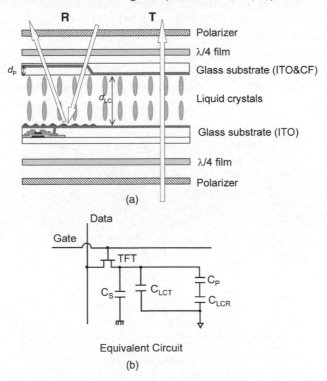

Figure 2.23 (a) Device configuration of a transflective LCD using a passivation layer for voltage shielding and (b) its equivalent circuit

d_P is formed between the front electrode and the LC cell with a thickness of d_{LC}. Electrically, this passivation layer functions as a capacitor, taking out part of the driving voltage from the two electrodes. From the equivalent circuit, the voltage V_D from the data line will be applied to the transmissive sub-pixel by $V_T = V_D$ and to the reflective sub-pixel by $V_R = C_P/(C_{LCR} + C_P) \cdot V_D$, where $C_{LCR} = \varepsilon_{LCR}/d_{LC}$ and $C_P = \varepsilon_P/d_P$. By tuning the dielectric constant ε_P and thickness d_P of the passivation layer, the VT and VR curves can be matched better.

However, from the voltage difference $\Delta V = (C_{LCR})/(C_{LCR} + C_P) \cdot V_D$ between the transmissive and reflective regions, the adjusted VR curve will be shifted along the voltage axis towards the higher voltage end. Referring back to Figure 2.16, a simple shift to the right of the VR curve will not make the VT and new VR curves match better because their slopes are different. Another issue is the difference in threshold voltage. From the voltage shift, the reflective mode will exhibit a higher threshold voltage than the transmissive mode, which is not preferred. To demonstrate these issues, in Figure 2.24 we plot the VT and VR curves of a 4 µm VA cell using MLC-6608, where the passivation layer is made of SiO_2 ($\varepsilon \sim 3.9$) with 1 µm thickness. Here, T stands for the VT curve of the transmissive sub-pixel, R1 stands for the VR curve from the reflective region if the passivation layer has zero thickness, and R2 stands for the VR curve with 1 µm SiO_2 passivation layer. We can see that the VR in R2 still

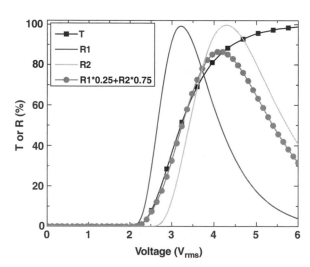

Figure 2.24 VT and VR curves of the voltage-shielding method using a passivation layer as a capacitor

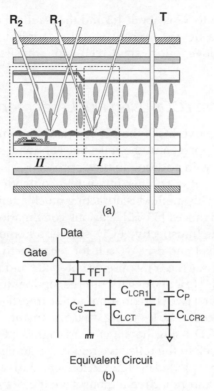

(a)

Equivalent Circuit
(b)

Figure 2.25 (a) Device configuration of a transflective LCD with two reflective regions and (b) its equivalent circuit

does not have good overlap with the VT curve, and the threshold voltage increases from ∼2.2 V_{rms} to ∼2.6 V_{rms}. Therefore, using a passivation layer in series with the reflective LC cell is not adequate to match the VT and VR curves, since their slopes are quite different.

To solve this problem, another approach, which is one step on from the voltage-shielding method, has been proposed. The device configuration is shown in Figure 2.25(a) and its equivalent circuit is plotted in Figure 2.25(b) [51, 52]. In this design, the reflective region is divided into two parts: region I with LCs sandwiched between the driving electrodes directly; and region II with an additional passivation layer inserted between the LCs and the driving electrodes. The VR curve for region I is designated R1 in Figure 2.24 and has the same threshold voltage as the VT curve; the VR curve for region II is called R2 in Figure 2.24. By adjusting the ratio between R1 and R2, we can obtain an averaged VR curve that exhibits a good match with the VT curve. The optimum ratio of R1 to R2 is about 1:3 and the resultant curve is also plotted in Figure 2.24. This device concept is quite simple to implement and is now widely employed for single-cell-gap transflective LCDs using vertical alignment.

2.4.4 Transflective LCDs Using Dual Alignment

For an ideal transflective LCD, the phase retardation in the bright state for the transmissive sub-pixel and the reflective sub-pixel should be about $\lambda/2$ and $\lambda/4$, respectively. In addition to the approaches discussed above, using dual surface alignment is another interesting contender for single-cell-gap transflective LCDs. This method is attracting much attention as photo-alignment technology advances [53–55]. Various combinations of LC alignment can be applied to the transflective LCD, such as a homogeneous cell for the transmissive sub-pixel and a HAN cell for the reflective sub-pixel [56–59], a homogeneous cell for the transmissive sub-pixel and an MTN cell for the reflective sub-pixel [41], or one of various other options [60–63]. In this section we will employ the first two combinations to illustrate the underlying physics in designing transflective LCDs using dual alignment.

A transflective LCD using homogeneous alignment in the transmissive region and a HAN cell in the reflective region is a straightforward case. The device configuration is plotted in Figure 2.26. Here, the rear surface alignments in the transmissive and reflective regions have the same pre-tilt angle ($\sim2^\circ$ to 5°), which can be obtained by mechanical rubbing or by photo-alignment.

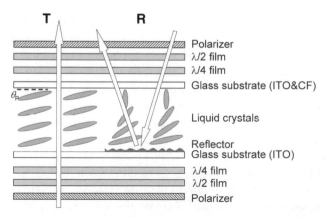

Figure 2.26 Device configuration of a transflective LCD using homogeneous alignment in the transmissive region and HAN alignment in the reflective region

On the front surface, for example, the vertical alignment may be coated first, and then the alignment layer in the transmissive region treated by UV or ion-beam exposure to obtain a homogeneous alignment with a surface pre-tilt angle at θ_P. Here, similar to the dual-cell-gap transflective LCD using homogeneous alignment, the residual surface phase retardation at a relatively high voltage also exists, requiring compensation from the quarter-wave plates to attain a good dark state. In other words, the optical axes of the quarter-wave plates need to be set either parallel or perpendicular to the LC rubbing direction. In addition, because the HAN cell usually exhibits a much lower threshold voltage than the homogeneous one, the match between VT and VR curves needs to be adjusted by reducing the threshold voltage of the transmissive sub-pixel, for example by increasing its pre-tilt angle on the front surface or by using an LC material with a large $\Delta\varepsilon$.

To study the electro-optics of this device, we performed a simulation of a 3.6 μm LC cell using MLC-6686 with $\Delta n \sim 0.097$ at $\lambda = 550$ nm. The rear surface rubbing angle is set at 3° for both homogeneous and HAN cells. The simulated VT and VR curves are plotted in Figure 2.27. As expected, if the front pre-tilt angle in the transmissive region is small ($\sim 3°$), VT exhibits a threshold voltage of $\sim 1\ V_{rms}$, departing significantly from the VR curve with a very small threshold voltage. By increasing the surface pre-tilt angle [59] to about 45° and adjusting the compensating retardation values of the

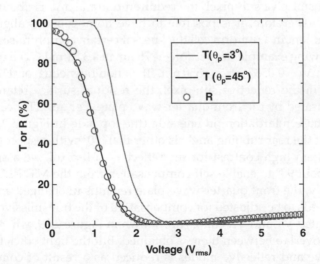

Figure 2.27 VT and VR curves for a transflective LCD using homogeneous and HAN alignment

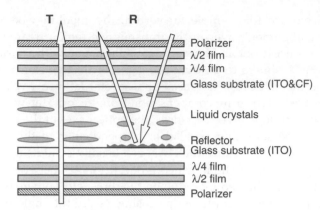

Figure 2.28 Device configuration of a transflective LCD using homogeneous alignment in the transmissive region and MTN alignment in the reflective region

quarter-wave plates, the threshold voltage of the VT curve drops and a good match can be obtained. Here, the elevated tail at higher voltages results from compensation from the quarter-wave plates.

In a second example of a transflective LCD using dual alignment, the HAN cell in the reflective sub-pixel is replaced by an MTN cell, as shown in Figure 2.28 [41]. For the reflective 70° and 90° MTN cells, the optimal phase retardation is typically about 278 nm and 240 nm, respectively [64]. Typically, for the transmissive sub-pixel, the required optimal $d\Delta n$ is about 300 nm at $\lambda = 550$ nm. Therefore, it is possible to have homogeneous alignment and MTN alignment in a uniform cell thickness if certain compromises are made. In the following example, we set $d\Delta n \sim 275$ nm in a 3.5 μm LC cell using MLC-6614 with $\Delta n \sim 0.0787$ at $\lambda = 550$ nm. If a homogeneous or HAN cell is employed in the reflective sub-pixel, the residual surface retardation can be compensated by the front quarter-wave plate. For an MTN cell, only the residual phase retardation on one side (the top side in Figure 2.28) can be cancelled if the rear rubbing angle is other than 90° with respect to the front one. To obtain high contrast for the reflective sub-pixel, we assign a twist angle of about 90° to enable self-compensation from the MTN LC cell itself. Accordingly, the front quarter-wave plate remains unchanged and only the rear one needs to be adjusted for compensation of the transmissive sub-pixel. The calculated VT and VR curves are shown in Figure 2.29, where a reasonably good overlap between them is obtained. But the light efficiency of both transmissive and reflective modes is reduced as a result of compromising on the $d\Delta n$ value. If contrast in the reflective sub-pixel could be sacrificed a little, a 70° MTN cell could be used for good light efficiency.

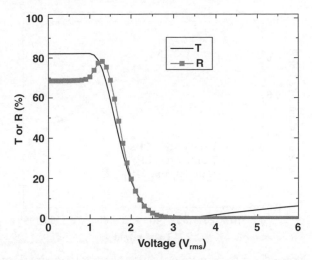

Figure 2.29 VT and VR curves for a transflective LCD using homogeneous and 90° MTN alignment

In the above examples, the reflective sub-pixel uses a HAN cell or an MTN cell, but the transmissive sub-pixel employs homogeneous alignment. We may consider replacing the homogeneous cell with a TN cell in the transmissive region. However, a TN cell has a relatively low transmittance between two circular polarizers [65]. Figure 2.30 plots the normalized light efficiency in a reflective MTN cell, a transmissive TN cell between two linear polarizers, and a transmissive TN cell between two circular polarizers as a function of the LC twist angle. Obviously, a pure phase retardation effect is preferred for the transmissive cell between two circular polarizers.

From the above studies, we find that device performance using dual alignment can be optimized for both good light efficiency and a single gamma driving in a single-cell-gap configuration. However, light leakage near the boundary of the transmissive and reflective regions could reduce the contrast ratio of the device because of a mismatch in the phase retardation of the LC layer (LC transition from one alignment to another) and the compensation quarter-wave plates there. This might require a black matrix to suppress light leakage. Similarly, the viewing angle of this display using homogeneous alignment is inadequate. To widen the viewing angle, IPS or FFS modes could be employed in the transmissive sub-pixel, and this will be further discussed in Chapter 4. Presently, the major obstacle is fabrication complexity, especially with regard to forming a distinctive alignment on one side of the substrate in each small pixel. Nevertheless, with advances in

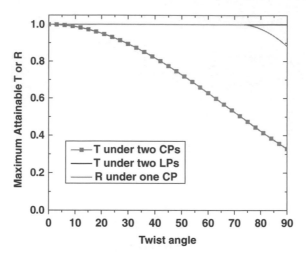

Figure 2.30 Normalized maximum possible light efficiency for a reflective MTN cell, a transmissive TN cell under crossed linear polarizers, and a transmissive TN cell under two circular polarizers with respect to the twist angle (redrawn from (65))

photo-alignment technology, dual-alignment technology certainly holds a great deal of promise for high-performance transflective LCDs.

2.5 Summary

In this chapter, numerical modeling methods for calculating LC director orientation and LC optics have been introduced. With regard to LC director modeling, the basic idea is to minimize the free energy in the system. Under different external conditions, such as constant voltage or constant charge, careful consideration should be given to the representation used for the free energy. Vector or tensor representation can be used, in accordance with the type of problem. FDM, FEM, or a combination of the two are useful numerical tools for minimizing the energy. At present, several commercial simulators have been developed with their own unique characteristics. For LC optics, both the 4×4 matrix method and the 2×2 extended Jones matrix method have been introduced and studied. The 4×4 matrix method is a complete solution to the 1D optics of systems with both uniaxial and biaxial media. For display applications, a local spectral average is usually needed to eliminate the impact of oscillations from internal reflections. The 2×2 extended Jones matrix is much simpler and faster in computing the LC optics, provided that internal reflections can be ignored. Thus, this method is extensively employed in modeling LCDs. Following the modeling sections, several numerical examples were given to illustrate their functions.

The introduction of several selected transflective LCD configurations will help readers to understand device physics, optics, and design considerations. Different methods targeting good light efficiency, such as the dual-cell-gap method, dual gamma curve method, dual field method, and dual alignment method, have been introduced. Specific examples were given in each part to study their electro-optics. The design considerations and methodology were also included in the description of each device. In addition to these typical methods, there are other configurations using different mechanisms, such as a transflective LCD using wire grid polarizers, but these have not been covered in this chapter. These topics will be addressed in Chapter 4.

We will continue with more advanced topics associated with transflective LCDs, such as wide viewing angle and weak color shift. The specific topics are mainly used in designing broadband and/or wide viewing angle circular polarizers for VA-based devices, or designing FFS-based transflective LCDs with or without in-cell retarders. All these topics will be discussed in subsequent chapters.

Appendix 2.A

In a homogeneous uniaxial medium layer with tilt angle θ and azimuthal angle ϕ, we can express the dielectric tensor as:

$$\overleftrightarrow{\varepsilon} = \begin{pmatrix} \varepsilon_{xx} & \varepsilon_{xy} & \varepsilon_{xz} \\ \varepsilon_{yx} & \varepsilon_{yy} & \varepsilon_{yz} \\ \varepsilon_{zx} & \varepsilon_{zy} & \varepsilon_{zz} \end{pmatrix}. \tag{2A.1}$$

where

$$\varepsilon_{xx} = n_o^2 + (n_e^2 - n_o^2)\cos^2\theta \cos^2\phi, \tag{2A.2a}$$

$$\varepsilon_{xy} = \varepsilon_{yx} = (n_e^2 - n_o^2)\cos^2\theta \sin\phi \cos\phi, \tag{2A.2b}$$

$$\varepsilon_{xz} = \varepsilon_{zx} = (n_e^2 - n_o^2)\sin\theta \cos\theta \cos\phi, \tag{2A.2c}$$

$$\varepsilon_{yy} = n_o^2 + (n_e^2 - n_o^2)\cos^2\theta \sin^2\phi, \tag{2A.2d}$$

$$\varepsilon_{yz} = \varepsilon_{zy} = (n_e^2 - n_o^2)\sin\theta\,\cos\theta\,\sin\phi, \tag{2A.2e}$$

$$\varepsilon_{zz} = n_o^2 + (n_e^2 - n_o^2)\sin^2\theta, \tag{2A.2f}$$

where n_o and n_e are the ordinary and extraordinary refractive indices of each medium layer, respectively. For absorption materials, such as a polarizer, the refractive indices are complex numbers. With dielectric tensor information, an extended Jones matrix for that layer can be specified.

For the ith sub-layer, its element matrix $\mathbf{J}_i$ is equal to:

$$\mathbf{J}_i = (\mathbf{SGS}^{-1})_i, \qquad (i = 1, 2, \ldots, N) \tag{2A.3}$$

Here,

$$\mathbf{S} = \begin{bmatrix} 1 & c_2 \\ c_1 & 1 \end{bmatrix} \tag{2A.4}$$

and

$$\mathbf{G} = \begin{bmatrix} \exp(ik_{z1}d_i) & 0 \\ 0 & \exp(ik_{z2}d_i) \end{bmatrix}, \tag{2A.5}$$

where d_i is the thickness of the corresponding ith layer, and

$$\frac{k_{z1}}{k_0} = \sqrt{\left[n_o^2 - \left(\frac{k_x}{k_0}\right)^2 \right]}, \tag{2A.6}$$

$$\frac{k_{z2}}{k_0} = -\frac{\varepsilon_{xz}}{\varepsilon_{zz}}\frac{k_x}{k_0} + \frac{n_o n_e}{\varepsilon_{zz}} \times \left(\left[\varepsilon_{zz} - \left(1 - \frac{n_e^2 - n_o^2}{n_e^2}\cos^2\theta\,\sin^2\phi\right)\left(\frac{k_x}{k_0}\right)^2 \right] \right)^{1/2}, \tag{2A.7}$$

$$c_1 = \frac{\left[\left(\frac{k_x}{k_0}\right)^2 - \varepsilon_{zz}\right]\varepsilon_{yx} + \left[\left(\frac{k_x}{k_0}\right)\left(\frac{k_{z1}}{k_0}\right) + \varepsilon_{zx}\right]\varepsilon_{yz}}{\left[\left(\frac{k_x}{k_0}\right)^2 + \left(\frac{k_{z1}}{k_0}\right)^2 - \varepsilon_{yy}\right]\left[\left(\frac{k_x}{k_0}\right)^2 - \varepsilon_{zz}\right] - \varepsilon_{yz}\varepsilon_{zy}}, \tag{2A.8}$$

$$c_2 = \frac{\left[\left(\frac{k_x}{k_0}\right)^2 - \varepsilon_{zz}\right]\varepsilon_{xy} + \left[\left(\frac{k_x}{k_0}\right)\left(\frac{k_{z2}}{k_0}\right) + \varepsilon_{xz}\right]\varepsilon_{zy}}{\left[\left(\frac{k_{z2}}{k_0}\right)^2 - \varepsilon_{xx}\right]\left[\left(\frac{k_x}{k_0}\right)^2 - \varepsilon_{zz}\right] - \left[\left(\frac{k_x}{k_0}\right)\left(\frac{k_{z2}}{k_0}\right) + \varepsilon_{zx}\right]\left[\left(\frac{k_x}{k_0}\right)\left(\frac{k_{z2}}{k_0}\right) + \varepsilon_{xz}\right]}$$

$$(2A.9)$$

Here $k_x = k_0 \sin\theta_k$ is the x component of the wave vector which is consistent in all layers, where $k_0 = 2\pi/\lambda$ and θ_k is the incident angle in air with respect to the $+z$-axis.

The analytical expressions cited above are limited to the uniaxial media in the system. For biaxial media, coordinated with n_z refractive index always along the z-axis in Figure 2.31, the tensor can be represented as:

$$\varepsilon_{xx} = n_x^2 + (n_y^2 - n_x^2)\sin^2\phi, \quad (2A.10a)$$

$$\varepsilon_{xy} = \varepsilon_{yx} = (n_x^2 - n_y^2)\sin\phi\,\cos\phi, \quad (2A.10b)$$

$$\varepsilon_{xz} = \varepsilon_{zx} = 0, \quad (2A.10c)$$

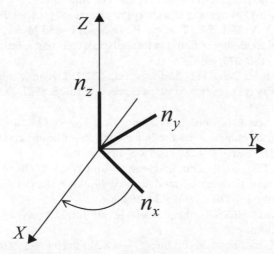

Figure 2.31 Representation of a biaxial medium such as a biaxial compensation film with refractive indices n_x, n_y, and n_z

$$\varepsilon_{yy} = n_y^2 + (n_x^2 - n_y^2)\sin^2\phi, \tag{2A.10d}$$

$$\varepsilon_{yz} = \varepsilon_{zy} = 0, \tag{2A.10e}$$

$$\varepsilon_{zz} = n_z^2, \tag{2A.10f}$$

We can always exchange n_x, n_y, and n_z in the above equations to suit different problems, as long as one refractive index is always along the z-axis. For a more general case, where the n_z refractive index has a non-zero angle with the z-axis, we can use the Euler transformation to obtain the expressions for each tensor component. Finally, the tensor components can be fed into derivations of the 4×4 or 2×2 matrix methods to model a display system with a biaxial compensation film.

References

[1] Anderson, J.E., Watson, P. and Bos, P.J. (1999) *LC3D: Liquid Crystal Display 3-D Directory Simulator, Software and Technology Guide.* Artech House Publishers.

[2] Thurston, R.N. and Berreman, D.W. (1981) Equilibrium and stability of liquid-crystal configurations in an electric field. *J. Appl. Phys.*, **52**, 508.

[3] Yang, D.K. and Wu, S.T. (2006) *Fundamentals of Liquid Crystal Devices.* John Wiley & Sons Ltd, Chichester.

[4] Davies, J.B., Day, S.E., Di Pasquale, F. and Fernandez, F.A. (1996) Finite-element modeling in 2-D of nematic liquid crystal structure. *Electron. Lett.*, **32**, 582–583.

[5] Fernandez, F.A., Day, S.E., Trwoga, P., Deng, H.F. and James, R. (2002) Three-dimensional modeling of liquid crystal display cells using finite elements. *Mol. Cryst. Liq. Cryst.*, **375**, 291–299.

[6] Fernandez, F.A., Deng, H.F. and Day, S.E. (2002) Dynamic modeling of liquid crystal display cells using a constant charge approach. *IEEE Trans. on Magn.*, **38**, 821–824.

[7] Yoon, H.J., Lee, J.H., Choi, M.W., Kim, J.W., Kwon, O.K. and Won, T. (2003) Comparison of numerical methods for analysis of liquid crystal cell: in-plane switching. *SID Symp. Dig.*, **50** (1), 1378–1381.

[8] Ge, Z., Wu, T.X., Lu, R., Zhu, X., Hong, Q. and Wu, S.T. (2005) Comprehensive three-dimensional dynamic modeling of liquid crystal devices using finite element method. *J. Disp. Technol.*, **1**, 194–206.

[9] Autronic-MELCHERS GmbH, available at: http://www.autronic-melchers.com/index.htm.

[10] SHINTECH. Inc., available at: http://www.shintech.jp/eng/index_e.html.

[11] SANAYI System Co., Ltd., available at: http://www.sanayisystem.com/introduction.html.

[12] Chigrinov, V.G., Kwok, H.S., Yakovlev, D.A., Simonenko, G.V. and Tsoy, V.I. (2004) LCD Optimization and Modeling. Invited paper, *SID Symp. Dig.*, **28** (1), 982–985.

[13] Chigrinov, V., Podyachev, Y., Simonenko, G. and Yakovlev, D. (2000) The Optimization of LCD Electrooptical Behavior using MOUSE-LCD Software. *Mol. Cryst. And Liq. Cryst.*, **351**, 17–25.

[14] LCQuest, available at: http://www.eng.ox.ac.uk/lcquest/.

[15] Jones, R.C., (1941) A new calculus for the treatment of optical systems. I. Description and discussion of the calculus. *J. Opt. Soc. Am.*, **31**, 488–500.

[16] Yeh, P., (1982) Extended Jones matrix method. *J. Opt. Soc. Am.*, **72**, 507–513.

[17] Gu, C. and Yeh, P. (1993) Extended Jones matrix method II. *J. Opt. Soc. Am. A*, **10**, 966–973.

[18] Lien, A., (1990) Extended Jones matrix representation for the twisted nematic liquid-crystal display at oblique incidence. *Appl. Phys. Lett.*, **57**, 2767–2769.

[19] Lien, A. and Chen, C.J. (1996) A new 2×2 matrix representation for twisted nematic liquid crystal displays at oblique incidence. *Jpn. J. Appl. Phys.*, **35**, L1200–L1203.

[20] Lien, A., (1997) A detailed derivation of extended Jones matrix representation for twisted nematic liquid crystal displays. *Liq. Cryst.*, **22**, 171–175.

[21] Chen, C.J., Lien, A. and Nathan, M.I. (1997) 4×4 and 2×2 matrix formulations for the optics in stratified and biaxial media. *J. Opt. Soc. Am.*, **14**, 3125–3133.

[22] Ge, Z., Wu, T.X., Zhu, X. and Wu, S.T. (2005) Reflective liquid-crystal displays with asymmetric incident and exit angles. *J. Opt. Soc. Am.*, **22**, 966–977.

[23] Berreman, D.W., (1972) Optics in stratified and anisotropic media: 4×4 matrix formulation. *J. Opt. Soc. Am.*, **62**, 502–510.

[24] Abdulhalim, I., Benguigui, L. and Weil, R. (1985) Selective reflection by helicoidal liquid crystals. Results of an exact calculation using the 4^*4 characteristic matrix method. *J. Physique*, **46**, 815–825.

[25] Wohler, H., Hass, G., Fritsch, M. and Mlynski, D.A. (1988) Faster 4×4 matrix method for uniaxial inhomogeneous media. *J. Opt. Soc. Am. A*, **5**, 1554–1557.

[26] Stallinga, S., (1999) Berreman 4×4 matrix method for reflective liquid crystal displays. *J. Appl. Phys.*, **85**, 3023–3031.

[27] Huang, Y., Wu, T.X. and Wu, S.T. (2003) Simulations of liquid-crystal Fabry–Perot etalons by an improved 4×4 matrix method. *J. Appl. Phys.*, **93**, 2490.

[28] De Gennes, P.G. and Prost, J. (1993) *The Physics of Liquid Crystals*, 2nd edition. Oxford Science Publications.

[29] Oseen, C.W., (1933) The theory of liquid crystals. *Trans. Faraday Soc.*, **29**, 883.

[30] Frank, F.C., (1958) On the theory of liquid crystals. *Discuss. Faraday Soc.*, **25**, 19.

[31] Rapini, A. and Papoular, M.J. (1969) Distortion d'une lamelle nématique sous champ magnétique conditions d'ancrage aus parois. *J. Phys. Colloq.*, **30**, C4.

[32] Davidson, A.J. and Mottram, N.J. (2002) Flexoelectric switching in bistable nematic device. *Phys. Rev. E*, **65**, 051710.

[33] Brown, C.V. and Mottram, N.J. (2003) Influence of flexoelectricity above the nematic Fréedericksz transition. *Phys. Rev. E*, **68**, 031702.

[34] Dickmann, S., Eschler, J., Cossalter, O. and Mlynski, D.A. (1993) Simulation of LCDs Including Elastic Anisotropy and Inhomogeneous Field. *SID Tech. Digest*, **24**, 638–641.

[35] Mori, H., Gartland, E.C. Jr., Kelly, J.R. and Bos, P.J. (1999) Multidimensional director modeling using the Q tensor representation in a liquid crystal cell and its application to the π cell with patterned electrodes. *Jpn. J. Appl. Phys.*, **38**, 135–146.

[36] Berreman, D.W. and Meiboom, S. (1984) Tensor representation of Oseen–Frank strain energy in uniaxial cholesterics. *Phys. Rev. A*, **30**, 1955–1959.

[37] Berreman, D.W. (1983) Numerical modeling of twist nematic devices. *Phil. Trans. R. Soc. Lond. A*, **309**, 203–216.

[38] Anderson, J.E., Titus, C., Watson, P. and Bos, P.J. (2000) Significant speed and stability increases in multi-dimensional director simulations. *SID Tech. Digest*, **31**, 906–909.

[39] Ting, D.L., Chang, W.C., Liu, C.Y., Shiu, J.W., Wen, C.J., Chao, C.H., Chuang, L.S. and Chang, C.C. (1999) A high brightness and high contrast reflective LCD with micro slant reflector (MSR). *SID Tech. Digest*, **30**, 954–957.

[40] Wu, S.T. and Wu, C.S. (1996) Mixed-mode twisted nematic liquid crystal cells for reflective displays. *Appl. Phys. Lett.*, **68**, 1455–1457.

[41] Okamoto, M., Hiraki, H. and Mitsui, S. (2001) Liquid crystal display. U.S. Patent 6281952, August.

[42] Zhu, X., Ge, Z., Wu, T.X. and Wu, S.T. (2005) Transflective liquid crystal displays. *J. Disp. Technol.*, **1**, 15–29.

[43] Hosaki, K., Uesaka, T., Nishimura, S. and Mazaki, H. (2006) Comparison of viewing angle performance of TN-LCD and ECB-LCD using hybrid-aligned nematic compensators. *SID Tech. Digest*, **37**, 721–724.

[44] Uesaka, T., Ikeda, S., Nishimura, S. and Mazaki, H. (2008) Viewing-angle compensation of TN- and ECB-LCD modes by using a rod-like liquid-crystalline polymer film. *J. Soc. Info. Disp.*, **16** (2), 257.

[45] Sheu, C.-R., Liu, K.-H., Hsin, L.-P., Fan, Y.-Y., Lin, I.-J., Chen, C.-C., Chang, B.-C., Chen, C.-Y. and Shen, Y.R. (2003) A Novel LTPS Transflective TFT LCD Driving by Double Gamma Method. *SID Tech. Digest*, **34**, 653–655.

[46] Lee, S.H., Do, H.W., Lee, G.D., Yoon, T.H. and Kim, J.C. (2004) A Novel Transflective Liquid Crystal Display with a Periodically Patterned Electrode. *Jpn. J. Appl. Phys.*, Part 2, **42**, L1455.

[47] Ge, Z., Zhu, X., Lu, R., Wu, T.X. and Wu, S.T. (2007) Transflective liquid crystal display using commonly biased reflectors. *Appl. Phys. Lett.*, **90**, 221111.

[48] Lu, R., Ge, Z. and Wu, S.T. (2008) Wide-view and single cell gap transflective liquid crystal display using slit-induced multidomain structures. *Appl. Phys. Lett.*, **92**, 191102.

[49] Roosendaal, S.J. (2003) Transflective liquid crystal display device. International Publication No. WO 03/048847 A1, June 12.

[50] Kang, S.-G., Kim, S.-H., Song, S.-C., Park, W.-S., Yi, C., Kim, C.-W. and Chung, K.-H. (2004) Development of a Novel Transflective Color LTPS-LCD with Cap-Divided VA-Mode. *SID Tech. Digest*, **35**, 31–33.

[51] Yang, Y.-C., Choi, J.Y., Kim, J., Han, M., Chang, J., Bae, J., Park, D.-J., Kim, S.I., Roh, N.-S., Kim, Y.-J., Hong, M. and Chung, K. (2006) Single Cell Gap Transflective Mode for Vertically Aligned Negative Nematic Liquid Crystals. *SID Tech. Digest*, **37**, 829–831.

[52] Lin, C.H., Chen, Y.R., Hsu, S.C., Chen, C.Y., Chang, C.M. and Lien, A. (2008) A novel advanced wide-view transflective display. *J. Disp. Technol.*, **4**, 123.

[53] Gibbons, W.M., Shannon, P.J., Sun, S.T. and Swetlin, B.J. (1991) Surface mediated alignment of nematic liquid crystals with polarized laser light. *Nature*, **351**, 49–50.

[54] Schadt, M., Schmitt, K., Kozenkov, V. and Chigrinov, V. (1992) Surface-Induced Parallel Alignment of Liquid Crystals by Linearly Polymerized Photopolymers. *Jpn. J. Appl. Phys.*, **31**, 2155–2164.

[55] Chigrinov, V.G., Kwok, H.-S., Hasebe, H., Takatsu, H. and Takada, H. (2008) Liquid-crystal photoaligning by azo dyes. *J. Soc. Inf. Disp.*, **16**, 897.

[56] Fan, Y.Y., Chiang, H.C., Ho, T.Y., Chen, Y.M., Hung, Y.C., Lin, I.J., Sheu, C.R., Wu, C.W., Chen, D.J., Wang, J.Y., Chang, B.C., Wong, Y.J. and Liu, K.H. (2004) A single-cell-gap transflective LCD. *SID Tech. Digest*, **35**, 647–649.

[57] Yu, C.-J., Kim, J., Kim, D.-W. and Lee, S.-D. (2004) A transflective liquid crystal display in a multimode configuration. *SID Tech. Digest*, **35**, 642–645.

[58] Yu, C.-J., Kim, D.-W. and Lee, S.-D. (2004) Multimode transflective liquid crystal display with a single cell gap using a self-masking process of photoalignment. *Appl. Phys. Lett.*, **85**, 5146.

[59] Lim, Y.J., Song, J.H. and Lee, S.H. (2005) Transflective liquid crystal display with single cell gap and single gamma curve. *Jpn. J. Appl. Phys.*, **44**, 3080–3081.

[60] Lee, S.H., Park, K.-H., Gwag, J.S., Yoon, T.-H. and Kim, J.C. (2003) A multimode-type transflective liquid crystal display using the hybrid-aligned nematic and parallel-rubbed vertically aligned modes. *Jpn. J. Appl. Phys.*, *Part 1*, **42**, 5127–5132.

[61] Kim, J., Yu, C.-J., Kim, D.-W. and Lee, S.-D. (2005) Reflective and transflective liquid crystal displays for low-power mobile applications. *Proc. of SPIE*, **5936**, 593603–1.

[62] Lee, S.-R., Jung, M.J., Park, K.-H., Yoon, T.-H. and Kim, J.-C. (2005) Design of a transflective LCD in the OCB mode. *SID Tech. Digest*, **36**, 734–737.

[63] Xu, P., Mak, H.-Y., Du, T., Chigrinov, V.G. and Kwok, H.S. (2008) Single-cell-gap transflective liquid-crystal display using double- and single-mode approaches. *J. Soc. Inf. Disp.*, **16**, 1157.

[64] Wu, S.T. and Yang, D.K. (2001) *Reflective Liquid Crystal Displays*. John Wiley & Sons Inc., New York.
[65] Roosendaal, S.J., van der Zande, B.M.I., Nieuwkerk, A.C., Renders, C.A., Osenga, J.T.M., Doornkamp, C., Peeters, E., Bruinink, J., van Haaren, J.A.M.M. and Takahashi, S. (2003) Novel high performance transflective LCD with a patterned retarder. *SID Tech. Digest*, **34**, 78–81.

3

Light Polarization and Wide Viewing Angle

3.1 Poincaré Sphere for Light Polarization in LCDs

Fundamentally, the LC cell together with two crossed linear polarizers functions as a light valve to control the light transmittance by external voltages. The light output transmittance from the front linear polarizer is controlled by the phase retardation of the LC cell and retardation films. In some ways, the optimization of an LCD for wide viewing angle is associated with controlling the light polarization change by designing proper LC cell and film parameters. Various numerical methods, such as the extended Jones matrix method and 4×4 matrix method, can be applied to compute the polarization change and light output which, in turn, optimize system parameters. However, a shortcoming of relying on purely numerical results is that it does not create an intuitive understanding or interpretation of the underlying physics or optics and thus might fall short of providing clear optimization directions. Instead, tracing a polarization trajectory on the Poincaré sphere is an approach that is quite intuitive and informative with regard to obtaining and interpreting the optimized results.

The Poincaré sphere representation is a very useful geometrical means of solving problems involving the propagation of polarized light through birefringent media (such as an LCD device) [1–3]. Polarized light can always be represented in many different ways, such as by direct electric field vectors $\mathbf{E}_x$ and $\mathbf{E}_y$, or by its long-axis azimuthal angle α and ellipticity angle β, or by the use of Stokes parameters S_1, S_2, and S_3 [4]. On the Poincaré sphere, elliptically polarized light, with its polarization state having long-axis azimuthal angle α and ellipticity angle β, can be denoted by the point **P** with longitude 2α and latitude 2β, as shown in Figure 3.1. Here the radius of the sphere is a unit length, and the Poincaré sphere adopted in this chapter only describes the polarization state of the light on the sphere surface, not the intensity [2, 3]. Accordingly, the Stokes parameters can be expressed by these two angles, as shown in the same figure. The long-axis azimuthal angle α of the elliptically polarized light is with respect to a reference axis in real x-y-z coordinates, where the polarization along that reference axis has $S_1 = +1$. It is very important for readers to understand this reference axis before defining different optical films and utilizing the Poincaré sphere to study polarization changes. Meanwhile, a uniaxial retardation A-film with its optical axis oriented at an angle γ from the reference axis in real x-y-z coordinates can be represented by point **A** on the Poincaré sphere, which is located at longitude 2γ on the equator (notice the angle on the Poincaré sphere is twice the real angle in x-y-z coordinates, because a polarization $[E_x = \cos(\gamma), E_y = \sin(\gamma)]$ at this angle γ has $S_1 = \cos(2\gamma)$, and $S_2 = \sin(2\gamma)$. Suppose the above-mentioned elliptically polarized light (polarization at point **P**) passes through the uniaxial film (optical axis at point **A**), the overall polarization change shown on the Poincaré sphere is equivalent to rotate the polarization along axis **OA** from point **P** to point **Q** by a spherical angle Γ, as shown in Figure 3.1.

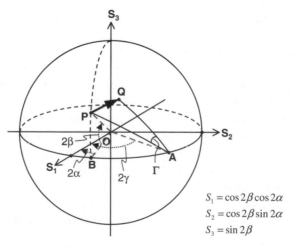

$$S_1 = \cos 2\beta \cos 2\alpha$$
$$S_2 = \cos 2\beta \sin 2\alpha$$
$$S_3 = \sin 2\beta$$

Figure 3.1 Light polarization change after passing a uniaxial film depicted on the Poincaré sphere

And arcs $\overline{PA}$ and $\overline{QA}$ are the spherical arcs on the great circle with a unit radius. Here Γ is determined by the off-axis phase retardation value of the uniaxial film, as discussed in Chapter 1. From the definition of a spherical triangle, the spherical angle is equal to the rotational angle. Here, we also need to note the difference between the real polarization trace **PQ** in this polarization rotation (shown in the figure) and the arc $\overline{PQ}$ used to determine the spherical triangle **PAQ** (not shown in the figure). The polarization trace **PQ** is obtained from rotation from **P** to **Q** on the Poincaré sphere along the axis **OA**, and it does not necessarily need to be on a great circle with a unit radius. In contrast, the arc $\overline{PQ}$ used for determining the spherical triangle **PAQ** is always defined as the arc connecting points **P** and **Q** on a large sphere with a unit radius. It should be pointed out that if the uniaxial film has a positive birefringence, then the above-mentioned rotation from point **P** to point **Q** is clockwise; the rotation is counterclockwise if the uniaxial film has a negative birefringence. Using these definitions, the detailed polarization state of light traversing through an LCD device can be depicted clearly on a Poincaré sphere.

Where the light is incident on the LCD system from an oblique angle, for a specific uniaxial film, a key step is first to define a reference axis on the incident wave plane and then the relative angles of all other retardation films and polarizers, as shown in Figure 3.2. For example, if one denotes the reference axis where $E_{//}$ is maximum and $S_1 = 1$ as **KJ**, then the next step is to determine the relative angles of other retardation films or polarizers such as **KM** or **KN** with respect to the reference axis **KJ** on the wave plane. Because the

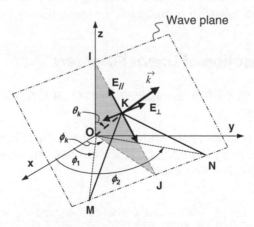

Figure 3.2 Wave incidence plane and the relation of two axes on the wave plane

azimuthal angles of axes **OJ**, **OM**, and **ON** in absolute x-y-z coordinates are known beforehand, we can then determine their relative angles with respect to the relative axis on the wave plane and in turn on the Poincaré sphere (x2). The equation used to calculate the relative angles on the incident wave plane from the known angles in absolute x-y-z coordinates is described in Chapter 1, and is repeated in the next section. Specifically, for a uniaxial A-film, when an observer views the LCD panel from an off-axis direction, its effective optical axis (assuming its optical axis is **OM** in x-y-z coordinates) on the wave plane is always correlated to the equator of the Poincaré sphere and varies as the viewing direction changes. But for a uniaxial C-film, its absolute optical axis is always along the z-axis in x-y-z coordinates. Thus, its projected axis on the wave plane is also easy to determine. As shown in Figure 3.2, if the $E_{//}$ polarization direction is chosen to be $S_1 = 1$, the optical axis of the C-film is also along this $S_1 = 1$ axis on the Poincaré sphere. On the other hand, if $E_{//}$ and $E_\perp$ shift their directions in the figure, i.e., to rotate $E_{//}$ by $90°$, the optical axis of the C-film is then along the direction $S_1 = -1$. With the reference axis having a polarization at $S_1 = 1$, we could further determine all the possible polarization states of the incident light by computing the Stokes parameters with the electric field components on the incident wave plane in Figure 3.2.

Based on these fundamental relations of light polarization changes and the associated optical axis definitions, we can draw the polarization trajectory on a Poincaré sphere for any film with known parameters. A good reference on this topic can be found in the work of Zhu *et al.* [3]. In the following sections we will apply this method to demonstrate some examples, for instance regarding compensating linear polarizers and designing wide-view circular polarizers that are critical elements in transflective or transmissive LCDs.

3.2 Compensation of Linear Polarizers

3.2.1 Deviation of the Effective Angle of Crossed Linear Polarizers

Based on the analysis and derivation in Chapter 1, the effective angle $\angle$**MKN** between the projected lines **KM** and **KN** viewed on the wave plane **MKN** in Figure 3.2 can be calculated from:

$$\varphi = \cos^{-1}\left[\frac{\cos(\phi_2 - \phi_1) - \sin^2\theta_k \cos(\phi_2 - \phi_k)\cos(\phi_1 - \phi_k)}{\sqrt{1 - \sin^2\theta_k \cos^2(\phi_2 - \phi_k)}\sqrt{1 - \sin^2\theta_k \cos^2(\phi_1 - \phi_k)}}\right], \quad (3.1)$$

where θ_k and ϕ_k are the polar angle and azimuthal angle in the medium of the wave vector $\vec{k}$. For incident light with a polar angle of θ_0 from air to the medium, the polar angle is equal to $\theta_k = \sin^{-1}(\sin\theta_0/n_p)$ in the medium with n_p (~ 1.5) defined as the real part of the average refractive index of the medium. In the special case of two crossed linear polarizers with $\phi_2 - \phi_1 = +90°$ (ϕ_1 and ϕ_2 are the absorption axes of the rear and front linear polarizers, respectively), the above equation can be written in another form in terms of $\phi_1 - \phi_k$ as:

$$\varphi = \cos^{-1}\left[\frac{1/2 \cdot \sin^2\theta_k \sin 2(\phi_1 - \phi_k)}{\sqrt{1 - \sin^2\theta_k + (1/2 \cdot \sin^2\theta_k \sin 2(\phi_1 - \phi_k))^2}}\right]$$

$$= \cos^{-1}\left[\frac{sign[\sin 2(\phi_1 - \phi_k)]}{\sqrt{(1 - \sin^2\theta_k)/(1/2 \cdot \sin^2\theta_k \sin 2(\phi_1 - \phi_k))^2 + 1}}\right]. \quad (3.2)$$

Therefore, if $\phi_1 - \phi_k = \pm 45°$, the absolute value of $(\varphi - 90°)$ reaches a maximum at $\left|\frac{\pi}{2} - \cos^{-1}\left(\frac{\sin^2\theta_0/n_p^2}{2 - \sin^2\theta_0/n_p^2}\right)\right|$, i.e., the effective angle of two polarizers has the largest deviation from the initial crossed configuration at 90°. Thus, the incident azimuthal directions at $\pm 45°$ away from the polarizer's transmission axis are defined as the bisector directions, where light leakage is most severe. On the other hand, when $\phi_1 - \phi_k = 0°$ or $\pm 90°$, the effective polarizer angle is always 90° regardless of the viewing polar angle, and light leakage is minimal in this direction if no compensation film is used. Therefore, for the compensation of the effective polarizer angle deviation, we simply focus on the bisector directions.

3.2.2 Compensation of Linear Polarizers using Uniaxial Films

Referring to Figure 3.2, we assume the polarizer absorption axes are set at $\phi_1 = 45°$ (rear) and $\phi_2 = -45°$ (front), and the incident angle is $\phi_k = -90°$ (at a bisector direction of $-45°$ from the rear polarizer's transmission axis) and incident polar angle θ_0 from air is 70° (with $\theta_k = \sin^{-1}(\sin\theta_0/n_p) = 38.78°$ in the medium). To make it easy to depict on the Poincaré sphere, we take the direction perpendicular to **KJ** (the incident direction) on the wave plane in Figure 3.2 ($E_{//}$ and $E_{\perp}$ need to be shifted in the plot) as the reference axis with

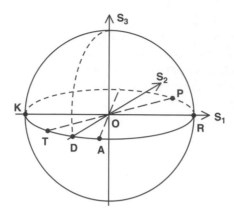

Figure 3.3 Angle representation with ϕ_1 (rear polarizer absorption axis) = 45° and ϕ_2 (front polarizer absorption axis) = −45°, incident angle ϕ_k = −90°, and the reference axis is set at ϕ_R = 0°

S_1 = +1 on the Poincaré sphere. In other words, along the incident direction **KJ** at ϕ_k = −90°, S_1 = −1. Consequently, the angle representations of the reference axis and the incident direction can be depicted as points **R** and **K** in Figure 3.3. Accordingly, the absorption axes of the rear and front polarizers are at points **P** and **A**, respectively, where points **P** and **A** are symmetrical with respect to the S_1 axis as $\angle$**AOR** = $\angle$**POR**. In addition, because the effective transmission axis of the rear linear polarizer is always orthogonal to its absorption axis at point **P**, an extension of point **P** along the sphere center **O** will determine the initial light polarization at point **T** on the Poincaré sphere when the light emerges from the rear linear polarizer. Therefore, using Equation (3.1), we can compute $\angle$**POA** $= 2\left[\cos^{-1}\left(\dfrac{\sin^2\theta_0/n_p^2}{2-\sin^2\theta_0/n_p^2}\right)\right] = 2\ \angle$**AOR** = $\sim 151.74°$, and $\angle$**AOR** = $\angle$**POR** = 75.87°. Without compensation, this deviation of effective polarizer angle (point **T** departs from point **A**) will cause light leakage at oblique incidence, especially in the bisector directions. The compensation mechanism takes the form of inserting a retardation film (or films) with the appropriate parameters between these two polarizers to adjust the final light polarization to point **A** on the Poincaré sphere, where the front linear polarizer absorption axis is located.

One compensation method uses a uniaxial positive C-film and a uniaxial positive A-film to adjust the final polarization to match the absorption axis of the front polarizer [5, 6]. The polarization change is traced and plotted in Figure 3.4, where the positive A-film has its optical axis parallel to the

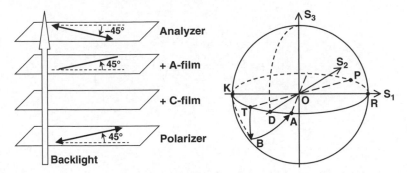

Figure 3.4 Film compensation of the two crossed linear polarizers and the corresponding polarization change on the Poincaré sphere

absorption axis of the rear linear polarizer. Note that, except in certain special situations, the compensation film generally should be aligned in such a way that there is zero phase retardation for the axially incident light in the LCD system, to eliminate γ its wavelength dispersion impact on the light. After incident light passes the rear linear polarizer, its polarization is at point **T**; the positive C-film (with its effective optical axis at point **K**) then rotates the polarization from point **T** to **B** along the **KO** axis. The phase retardation of the C-film determines the position of point **B**, which is closely related to the next rotation trace from point **B** to **A** along axis **PO** by the following positive A-film. In other words, the phase retardations of the uniaxial C-film and A-film are not independent.

Following the work of Zhu *et al.* [3], the retardation values of these films can be analytically determined based on trigonometry theory. First, in the spherical triangle **TAB** (note: the arc $\overline{\text{TB}}$ defined on a large sphere is not shown and is different from the polarization trace **TB** in the plot, and the same is true for the arc $\overline{\text{BA}}$), arc $\overline{\text{TB}}$ = arc $\overline{\text{TA}}$. Therefore, arc $\overline{\text{TB}}$ is known since arc $\overline{\text{TA}}$ is equal to $\angle$**TOA** at $\pi - 2\angle$**AOR**, where $\angle$**AOR** can be obtained from the previous analysis regarding the effective polarizer angle. In addition, in the spherical triangle **KTB**, arc $\overline{\text{KT}}$ = arc $\overline{\text{KB}}$ = $\angle$**KOT** = $\angle$**AOR**. By now, we have all three arc lengths of the spherical triangle **KTB**. From trigonometry theory, we can then obtain the spherical angle $\angle$**TKB**, which is equal to the phase retardation value of the C-film in this incident direction. If we designate $\angle$**AOR** as φ_0, then we can obtain $\angle$**TKB** = $2\sin^{-1}(ctg\varphi_0)$ and $\angle$**KTB** = $2\cos^{-1}\left(\frac{1}{\sqrt{2}\sin\varphi_0}\right) = \cos^{-1}(ctg^2\varphi_0)$. Thus, $\angle$**ATB** = $\pi - \angle$**TKB** = $\pi - \cos^{-1}(ctg^2\varphi_0)$. Finally, with the retardation values of the films, we can obtain the desired film thicknesses as:

$$d_c = \lambda \frac{\sin^{-1}(ctg(\varphi_0))/\pi}{\left[n_{o,c}\sqrt{1 - \frac{\sin^2\theta_0}{n_{e,c}^2}} - n_o\sqrt{1 - \frac{\sin^2\theta_0}{n_{o,c}^2}} \right]},$$ (3.3)

and

$$d_a = \lambda \frac{1/2 - \cos^{-1}(ctg^2(\varphi_0))/(2\pi)}{\left[n_{e,a}\sqrt{1 - \frac{\sin^2\theta_0}{2n_{e,a}^2} - \frac{\sin^2\theta_0}{2n_{o,a}^2}} - n_{o,a}\sqrt{1 - \frac{\sin^2\theta_0}{n_{o,a}^2}} \right]}.$$ (3.4)

To verify the derivations above, we calculate the angular light leakage of two crossed linear polarizers with and without compensation using A- and C-films. The results are shown in Figure 3.5. The polarizer and film parameters are set as: $n_{e\_pol} = 1.5 + i \times 2.208 \times 10^{-3}$, $n_{o\_pol} = 1.5 + i \times 3.222 \times 10^{-5}$, $n_{e\_a} = 1.5110$, $n_{o\_a} = 1.5095$, $n_{e\_c} = 1.5110$, and $n_{o\_c} = 1.5095$ at $\lambda = 550$ nm. If the compensation is targeted at $\theta_0 = 70°$, the calculated thicknesses of the C-film and A-film based on the above equations are $d_c = 60.09$ μm and $d_a = 92.59$ μm, respectively. For simplicity, the protective TAC film is not included in the calculation. To display the values relative to each other, the angular light leakage in the plot is normalized to the maximum axial transmittance from two parallel polarizers. Without compensation, the routine deviation of effective polarizer angle leads to 1% light leakage at about 40° (Figure 3.5(a)), but with compensation, the maximum light leakage from all angles is suppressed to ∼ 1.2×10^{-3} (Figure 3.5(b)).

In addition to using uniaxial positive C-film and positive A-film, other uniaxial film combinations can be applied, such as using a uniaxial negative A-film and a uniaxial positive A-film. Readers can find a more comprehensive discussion of such compensation combinations in [3]. Nevertheless, the analytical solution provides a powerful and intuitive method to design and optimize compensation films, but owing to the actual refractive index difference between adjacent layers and the fact that real polarizers are non-ideal, the optimal parameters might deviate a little from those obtained by the analytical methods discussed above.

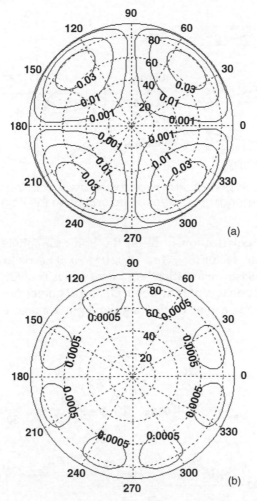

Figure 3.5 Angular light leakage from two crossed linear polarizers (a) without compensation and (b) with uniaxial A- and C-films (results are normalized to maximum parallel polarizer transmittance)

3.2.3 Compensation of Linear Polarizers using Biaxial Films

To convert the final polarization from point **T** to point **A**, at least two uniaxial films are required, but the same result could be achieved by using a single biaxial film [5, 7]. Figure 3.6 shows the compensation configuration

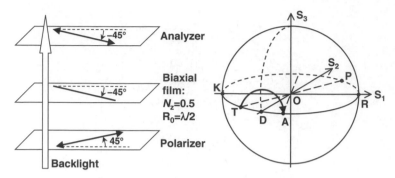

Figure 3.6 Film compensation of two crossed linear polarizers using one biaxial film and its corresponding polarization trajectory on the Poincaré sphere

and the related polarization trace on the Poincaré sphere. For this single biaxial film, the N_z factor ($N_z = (n_x - n_z)/(n_x - n_y)$) needs to be set at 0.5 and the in-plane phase retardation $R_0 = d \times (n_x - n_y) = \lambda/2$. The N_z factor controls the rotational center at point **D**, and R_0 determines the rotational angle. Figure 3.7 is a plot of the angular light leakage of the display. It

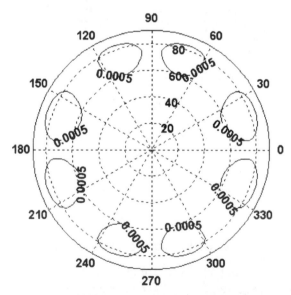

Figure 3.7 Angular light leakage from two crossed linear polarizers using one biaxial film at $\lambda = 550$ nm (normalized to maximum parallel polarizer transmittance)

shows that the maximum normalized light leakage is below 1.1×10^{-3}, which is similar to the above example using a uniaxial A-film and a uniaxial C-film. This simple compensation scheme is widely adopted nowadays for both IPS/FFS and MVA LCDs, owing to the well-developed fabrication technique for a half-wave biaxial film together with a linear sheet polarizer.

A shortcoming of using single biaxial film compensation is that it is difficult to achieve good compensation over a broad wavelength range. This can be clearly seen by reviewing the key film parameters – the N_z factor ($N_z = (n_x - n_z)/(n_x - n_y)$) and the in-plane phase retardation $R_0 = d \times (n_x - n_y)$. When the wavelength varies, $(n_x - n_z)$ and $(n_x - n_y)$ experience a similar dispersion tendency since both can be roughly characterized by the Cauchy relation as $\sim G \cdot \lambda^2 \lambda^{*2} / (\lambda^2 - \lambda^{*2})$ [8, 9]. Because the same bulk polymer material is used and treated to generate birefringence $(n_x - n_z)$ and $(n_x - n_y)$, their resonance absorption wavelength λ^* may be similar but G is different. In other words, the N_z factor may be roughly constant over most wavelengths of interest, leading to a roughly fixed rotational center at point **D** on the Poincaré sphere, but the in-plane retardation $R_0 = d \times (n_x - n_y)$ varies quite significantly for different wavelengths. For instance, a shorter wavelength leads to a larger rotational arc length, making the final polarization well behind point **A** on the Poincaré sphere in Figure 3.6. To overcome this problem, one solution is to develop a broadband half-wave biaxial film with inverted birefringence dispersion behavior by stacking two orthogonal films with the same N_z factor but different in-plane retardation $R_0 = d \times (n_x - n_y)$ and different wavelength-dependent birefringence slopes [10, 11], as discussed in Chapter 1. Another way is to use two biaxial films with different N_z factors but the same in-plane retardation $R_0 = d \times (n_x - n_y)$ [12]. The optical configuration of the second method is shown in Figure 3.8. The first half-wave biaxial film rotates the polarization from point **T** to point **D**, which is subsequently converted by the second film to the final point **A**. The corresponding angular light leakage at $\lambda = 550$ nm is shown in Figure 3.9, and the maximum value is significantly less than 6×10^{-4} (normalized value). (Compare with the above two examples.)

Trace $\overline{TD}$ and trace $\overline{DA}$ from different biaxial films could self-compensate for the wavelength dispersion. For example, at a shorter wavelength than 550 nm (the polarization trace is illustrated in Figure 3.10), both traces become longer, thus the trace from the first biaxial film will end at point **E** behind point **D**, and the longer trace from the second biaxial film then converts the final polarization at point **F**, which is still quite close to point **A**. Similar

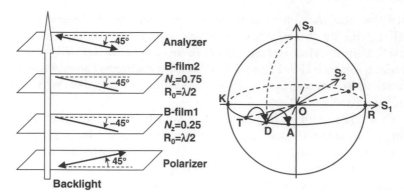

Figure 3.8 Film compensation of two crossed linear polarizers using two biaxial films and the corresponding polarization trajectories on the Poincaré sphere

compensation exists for longer wavelengths, where the trace point **E** is located in front of point **D**. Figure 3.11 shows the spectral light leakage from two crossed linear polarizers at off-axis incidence (with polar angle $\theta = 70°$ in the bisector direction) by using different compensation schemes. The light

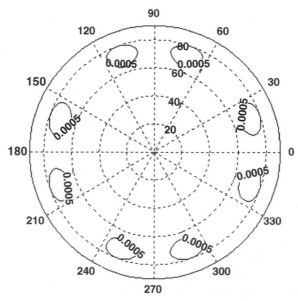

Figure 3.9 Angular light leakage from two crossed linear polarizers using two biaxial films at $\lambda = 550$ nm (normalized to maximum parallel polarizer transmittance)

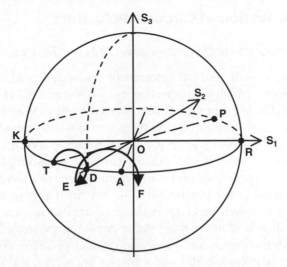

Figure 3.10 The polarization change at a shorter wavelength on the Poincaré sphere

leakage shown in the plot is the absolute value without normalization. Clearly, the display with compensation using two biaxial films exhibits much better broadband operation than the one using a single half-wave biaxial film or using uniaxial A- and C-films.

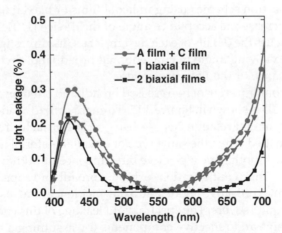

Figure 3.11 Spectral light leakage from two crossed linear polarizers using different film compensation with a polar angle of 70° in the bisector direction (without normalization)

3.3 Compensation of Circular Polarizers

3.3.1 Broadband and Wide-view Circular Polarizers

In this section we will study the design of wide-view and/or broadband circular polarizers, which are important optical components in MVA-based transflective LCDs. MVA mode with compensation films is a major candidate for wide-view (gray-scale inversion-free) transmissive/transflective LCDs for mobile displays [13–17]. Even for a transmissive MVA LCD for mobile devices, circular polarizers can greatly enhance transmittance, since the azimuthal angle dependence on LC directors is removed (compare the case using two crossed linear polarizers) [13]. Yet the viewing angle of most transmissive or transflective LCDs using circular polarizers is fairly narrow because of the increased number of films in circular polarizers. In this section, the compensation schemes for broadband circular polarizers consisting of a linear polarizer, a half-wave film, and a quarter-wave film, and a conventional one with a linear polarizer and a single quarter-wave plate, will be discussed [18–27].

A conventional cost-effective broadband circular polarizer consists of a linear polarizer, a uniaxial monochromatic half-wave film, and a uniaxial monochromatic quarter-wave film, with their optical axes aligned at certain angles [18]. To reduce cost, the retardation films are usually made of the same type of uniaxial positive A plates. However, the polarization produced is circular only at normal incidence, while its off-axis light leakage is significant. Many compensation schemes using uniaxial films or biaxial films have been proposed to increase the acceptance angle of the device [1, 21, 22]. Before we move on to discuss the details of each design, let us first investigate the origins of the narrow viewing angle in conventional broadband circular polarizers with the assistance of the Poincaré sphere.

The device configuration of two crossed broadband circular polarizers for a transflective LCD is shown in Figure 3.12. To enable broadband operation, the optical axis alignment relation $2\varphi_{1/4\lambda} - 4\varphi_{1/2\lambda} = 90°$ in each circular polarizer needs to be satisfied when the same type of uniaxial material is used for both retardation films (both have a positive birefringence or both have a negative birefringence). This configuration exhibits broadband operation for the transmissive and reflective modes at axial incidence, but its off-axis light leakage is still quite severe. The spectral light leakage of this configuration for both transmissive and reflective components (by assuming a metal reflector behind the first circular polarizer) is shown in Figure 3.13. At normal incidence, both transmissive and reflective modes exhibit a good dark state

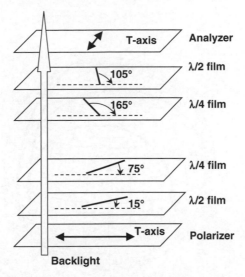

Figure 3.12 Conventional broadband circular polarizers using a linear polarizer, a half-wave film, and a quarter-wave film

covering a wide wavelength range of visible light. At off-axis incidence with $\varphi = -45°$ and $\theta = 45°$, the light leakage for both transmissive and reflective sub-pixels increases dramatically, indicating a very narrow viewing angle of the device. Here the light leakage values also vary with the refractive index values of retardation films.

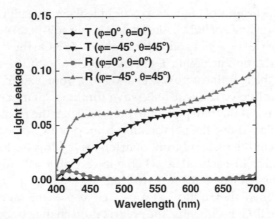

Figure 3.13 Simulated spectral light leakage of both transmissive and reflective sub-pixels (without normalization) of the circular polarizers shown in Figure 3.12

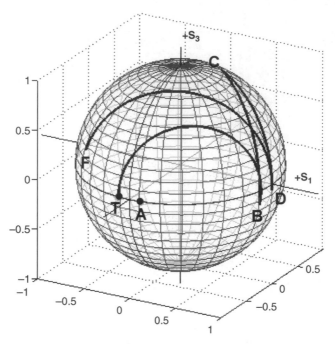

Figure 3.14 Polarization change of the transmissive sub-pixel on the Poincaré sphere incident at $\theta = 45°$ and $\phi = -45°$

To analyze the origins of high off-axis light leakage, we plot, in Figure 3.14, the polarization change of light in the transmissive sub-pixel when incident at an azimuthal angle $\varphi = -45°$, and polar angle $\theta = 45°$. On the Poincaré sphere, the linear polarization at points **T** and **A** coincide with the effective transmission axis of the rear linear polarizer and absorption axis of the front linear polarizer, respectively. The rear half-wave film and quarter-wave film first convert the polarization from point **T** to point **C**. Then the front half-wave and quarter-wave films move the polarization from point **C** to point **F**. Although the optical axes of the rear and front quarter-wave films (or half-wave films) are perpendicular to each other when viewed at $\theta = 0°$ (or at directions regardless of θ such as $\varphi = 15°$ for two half-wave films, or $\varphi = 75°$ for two quarter-wave films), their optical axes viewed at $\varphi = -45°$ and $\theta = 45°$ are no longer orthogonal. Consequently, the polarization change traces from the rear circular polarizer films and the front ones diverge from each other. Even at $\theta = 45°$, the final polarization just before the front linear polarizer at point **F** is quite far from the absorption axis at point **A**, indicating incomplete absorption of the light in this direction. The corresponding angular light leakage for the transmissive and reflective sub-pixels at $\lambda = 550$ nm is shown in Figure 3.15. From both transmissive and reflective sub-pixels, the 10% light leakage (normalized to the parallel polarizer transmittance) occurs at 40° in the polar direction. In summary, the origins of light leakage from conventional broadband circular polarizers are: (i) deviation of effective polarizer

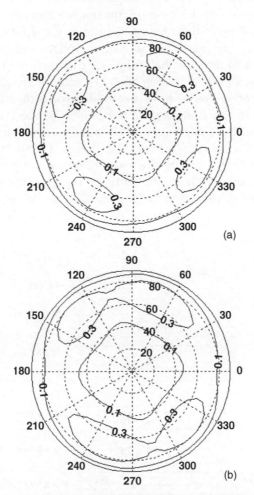

Figure 3.15 The simulated angular light leakage at $\lambda = 550$ nm (normalized to parallel polarizer transmittance) for (a) the transmissive sub-pixel and (b) the reflective sub-pixel of a transflective LCD using conventional polarizers

angle at different off-axis directions, i.e., point **T** departs from point **A**, and (ii) accumulation of phase retardation from multiple uniaxial quarter-wave and half-wave plates, i.e., final polarization at point **F** departs from starting polarization at point **T**. Therefore, to suppress light leakage and widen the viewing angle of circular polarizers, we should try to solve these two problems.

First we introduce a new broadband circular polarizer where retardation films from two circular polarizers can compensate each other to achieve a wide viewing angle. The device configuration is shown in Figure 3.16 [25], where the two half-wave films (and two quarter-wave films) from different circular polarizers are made of uniaxial materials with opposite optical birefringence. In addition, two half-wave (or quarter-wave) films have their optical axes aligned parallel to each other. To achieve broadband operation, the light polarization traces from the half-wave film and quarter-wave film need to be in the same top or bottom hemisphere on the Poincaré sphere. Therefore, the optical axis angle relation needs to be adjusted to $2\varphi_{1/4\lambda} - 4\varphi_{1/2\lambda} = -90° + m\pi$, where m is an integer. The spectral light leakage of the device for both transmissive and reflective sub-pixels is shown in Figure 3.17. Compared with the conventional design using the same type of uniaxial film, the improvement in the transmissive sub-pixel here is quite

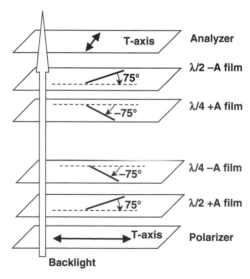

Figure 3.16 Broadband circular polarizers using positive and negative A-films

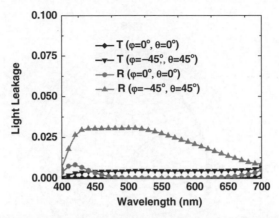

Figure 3.17 Simulated spectral light leakage of both transmissive and reflective modes (without normalization) for the circular polarizers shown in Figure 3.16

evident and the reflective sub-pixel also benefits from the self-compensation due to having the opposite birefringence. Please note that the optical axis arrangement of the positive or negative films could be in many other forms besides the one shown in Figure 3.16 [25].

Figure 3.18 depicts the light polarization change of the transmissive sub-pixel when light is incident from $\theta = 70°$ and $\phi = -45°$. A unique difference of this configuration from the conventional design using all positive films is that the half-wave (or quarter-wave) films from different circular polarizers always have their optical axes parallel to each other. As a result, the optical trace from the rear half-wave and quarter-wave films (from point **T** to point **C**) overlaps with that from the front films (from point **C** to point **F**), regardless of the incident azimuthal and polar angles. The small deviation of these two traces in the figure comes from the refractive index mismatch between the positive and negative uniaxial films used in the simulation. By eliminating the polarization trace separation between rear and front retardation films, the light leakage from this design merely originates from the angular deviation of the two linear polarizers, which could be compensated by a biaxial film as discussed in Section 3.2.3.

Figure 3.19 plots the corresponding angular light leakage at $\lambda = 550$ nm. For the transmissive sub-pixel, the maximum normalized light leakage is about 4×10^{-2} and 1% light leakage occurs at about 40° in the polar direction. As we can see, this light leakage level is quite close to that from the two crossed linear polarizers in Figure 3.5. For the reflective sub-pixel, 10% normalized light

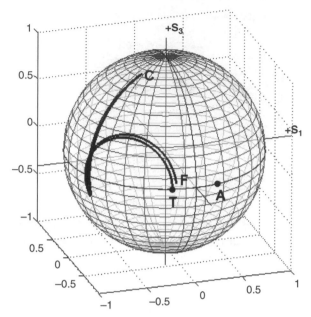

Figure 3.18 The polarization change on the Poincaré sphere for a transmissive cell with light incident from $\theta = 70°$ and $\phi = -45°$

leakage occurs at 50°, which is much better than that using all positive uniaxial films (shown in Figure 3.15(b)).

From Figure 3.18, we can see that an additional half-wave biaxial film between the front linear polarizer and the front half-wave film could draw the final polarization from point **F** closer to point **A**, similar to Figure 3.6. But the increased number of films would undoubtedly increase the device thickness and cost. Considering the real application of this circular polarizer in MVA transmissive or transflective LCDs, a negative C-film is always required to compensate the LC cell's phase retardation at an oblique angle of incidence. Hence, without adding an additional biaxial film, we can also use a negative uniaxial C-film to further widen the viewing angle to an acceptable level for mobile displays. The new device configuration is shown in Figure 3.20. Here, the optical axes of the half-wave and quarter-wave films are tuned away by a few degrees (but broadband operation in each circular polarizer is maintained) and their absolute phase retardations are intentionally adjusted downwards to 250 nm and 125 nm at $\lambda = 550$ nm, respectively. The negative C-film has its phase retardation at about -45 nm, which can actually be

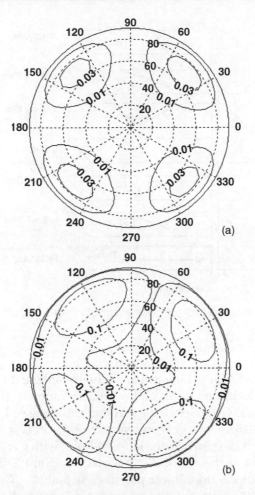

Figure 3.19 Simulated angular light leakage at $\lambda = 550$ nm (normalized to the parallel polarizer transmittance) of (a) the transmissive sub-pixel and (b) the reflective sub-pixel of the design shown in Figure 3.16

realized by the overall phase retardation from a negative C-film and a vertically aligned LC cell (which behaves like a positive C-film) in a real MVA LCD. With the above adjustment, the maximum normalized angular light leakage could be further suppressed to less than 2×10^{-2} and the angle for 1% light leakage could be extended to about 60° in the polar direction for the transmissive sub-pixel in Figure 3.21(a), yet the improvement for the reflective sub-pixel is limited, as depicted in Figure 3.21(b).

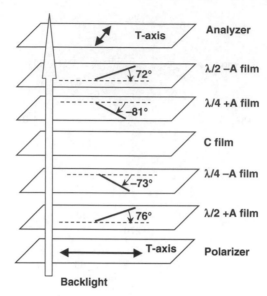

Figure 3.20 Broadband circular polarizers including a negative C-film to widen the viewing angle

These optimized film parameters are actually obtained by observing and adjusting the polarization traces on the Poincaré sphere. The compensation principle of the above device configuration can be viewed from the following polarization traces for light incident from $\theta = 70°$, $\varphi = -45°$ and $\theta = 70°$, $\varphi = 0°$, as Figure 3.22 shows. In the left-hand plot with $\theta = 70°$, $\varphi = -45°$, the transmission axis of the rear linear polarizer at point **T** departs from the absorption axis of the front linear polarizer at point **A**. The rear half-wave film first moves the polarization from point **T** to point **B**, which is further rotated to point **C** by the following rear quarter-wave film. The negative C-film converts the polarization further to point **D**, intentionally making the trace from the front films depart from the rear ones in contrast to that shown in Figure 3.18. Therefore, the final polarization from the front half-wave film is at point **F**, which is much closer to point **A** than that in Figure 3.18. Although by increasing the retardation value of this C-film, the final polarization **F** could even overlap with point **A**, it would conflict with the compensation in other directions. In the right-hand plot of the figure when viewed at $\theta = 70°$, $\varphi = 0°$, a larger trace **CD** from the C-film will move the final polarization at point **F** farther away from the front polarizer absorption

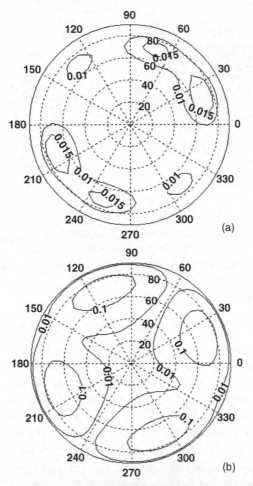

Figure 3.21 Calculated angular light leakage (normalized to the maximum transmittance of a pair of parallel polarizers) at $\lambda = 550\,nm$ for (a) the transmissive sub-pixel and (b) the reflective sub-pixel

axis at point **A**. Therefore, there is a compromise to be made when designing compensation film parameters between these two directions. Using only a negative C-film (along with the VA cell for actual applications) for compensation, a contrast ratio greater than 10:1 could be extended to greater than 70° in an MVA LCD [25], which is adequate for small-panel mobile displays.

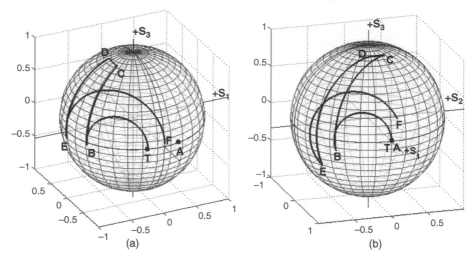

Figure 3.22 Polarization state trace on the Poincaré sphere when viewed from $\theta = 70°$ and (a) $\varphi = -45°$ and (b) $\varphi = 0°$

3.3.2 Narrow-band and Wide-view Circular Polarizers

In the configurations discussed above, each circular polarizer consists of at least a half-wave plate and a quarter-wave plate to enable broadband operation for both transmissive and reflective LCDs. Yet in some applications, the performance requirement of the reflective sub-pixel is less stringent, so that the half-wave plate can be removed. As a result, fewer films are used, thereby reducing the cost and thickness of mobile displays. As usual, to better understand the underlying compensation principles for circular polarizers with quarter-wave plates only, we first investigate the optics and light polarization associated with the conventional configuration comprising only a linear polarizer and a monochromatic quarter-wave plate. Figure 3.23 shows the optical configuration of this simple circular polarizer, while the spectral light leakage for both transmissive and reflective sub-pixels is shown in Figure 3.24. The transmissive sub-pixel exhibits low-level light leakage over a broad spectral range at normal incidence, but its off-axis light leakage is still quite significant. For the reflective sub-pixel (assuming a mirror is placed behind the front circular polarizer), a good dark state only occurs at the designed central wavelength (e.g., 550 nm) at on-axis incidence. High levels of light leakage appear when the wavelength deviates from the central value optimized for quarter-wave retardation of the film. In a similar way, as the incident angle increases, circular polarization cannot be conserved and light

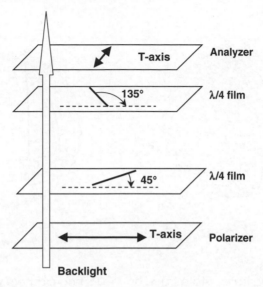

Figure 3.23 A simple circular polarizer consisting of a linear polarizer and a quarter-wave film

leakage becomes severe. Thus, for a reflective LCD, this simple circular polarization is definitely a narrow-band and narrow-view configuration.

The associated polarization traces for light incident from $\theta = 70°$, $\varphi = -45°$ and $\theta = 70°$, $\varphi = 0°$ are plotted in Figure 3.25. In the left plot with $\theta = 70°$, $\varphi = -45°$, the two quarter-wave films are orthogonal to each

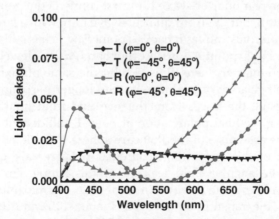

Figure 3.24 Calculated spectral light leakage for both T and R sub-pixels (without normalization) of the simple circular polarizer shown in Figure 3.23

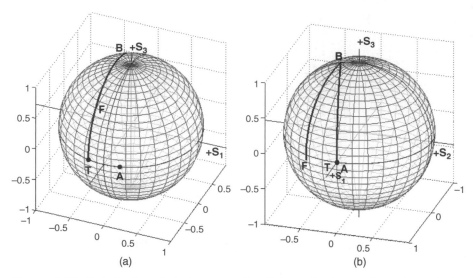

Figure 3.25 Polarization state trace on the Poincaré sphere when viewed at $\theta = 70°$ and (a) $\varphi = -45°$ and (b) $\varphi = 0°$

other, regardless of the incident polar angle. But their phase retardation values at an off-axis incident angle are not equal ($\Gamma_a = \frac{2\pi}{\lambda}d$ $\left[n_e \sqrt{1 - \frac{\sin^2\theta_0 \sin^2\phi_n}{n_e^2} - \frac{\sin^2\theta_0 \cos^2\phi_n}{n_o^2}} - n_o \sqrt{1 - \frac{\sin^2\theta_0}{n_o^2}} \right]$, their ϕ_n values, defined as the angle between the optical axis and the incident direction, are different; see Chapter 1). The rear quarter-wave film first converts the polarization from point **T** to point **B**, but the front quarter-wave film, with a smaller effective retardation value, only moves it back to point **F**, which is still far away from the absorption axis at point **A**. On the other hand, when viewed from $\theta = 70°$, $\varphi = 0°$), the two quarter-wave films exhibit the same off-axis phase retardation but their relative optical axes are no longer perpendicular to each other. As a result, the traces from the rear and front quarter-wave films diverge, making the final polarization at point **F** still distant from point **A**. Hence, with the assistance of the Poincaré sphere, the origins of the narrow viewing angle of this conventional circular polarizer are quite intuitive. Accordingly, the angular light leakage of this simple circular polarizer configuration at $\lambda = 550$ nm is shown in Figure 3.26. The maximum normalized light leakage of the transmissive sub-pixel is about 0.16, and 10% normalized light leakage occurs at angles greater than 50° in all directions. The reflective sub-pixel has similar angular light leakage to the transmissive sub-pixel.

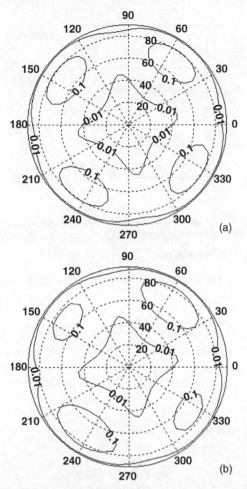

Figure 3.26 Calculated angular light leakage (normalized to maximum parallel polarizer transmittance) at $\lambda = 550\,\text{nm}$ for (a) the transmissive sub-pixel and (b) the reflective sub-pixel of the circular polarizers shown in Figure 3.23

To widen the viewing angle of the above circular polarizer configuration, we can insert a negative uniaxial (or biaxial) quarter-wave film in the rear side and a positive uniaxial quarter-wave film in the front side [23, 24]. The optical configuration is shown in Figure 3.27 and its spectral light leakage is plotted in Figure 3.28. The additional biaxial film could be a half-wave retardation film to compensate the polarizer´s effective angular deviation. Along with the

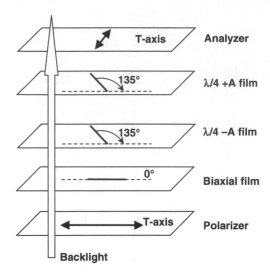

Figure 3.27 Circular polarizers using positive and negative uniaxial quarter-wave films

self-compensation from the two quarter-wave films, the transmissive sub-pixel could be fully compensated. In contrast to the examples discussed above, the light leakage of the transmissive sub-pixel from this design at $\theta = 45°$ almost coincides with that from the normal direction. Nevertheless,

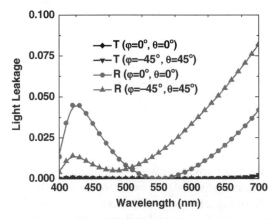

Figure 3.28 Calculated spectral light leakage for both transmissive and reflective sub-pixels (without normalization) for the circular polarizer configuration shown in Figure 3.27

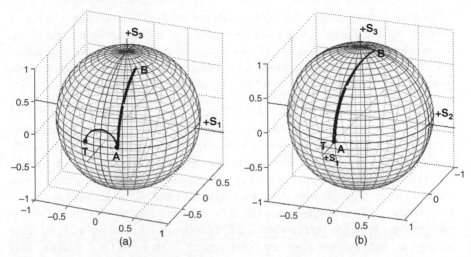

Figure 3.29 (a) Polarization state trace on the Poincaré sphere when viewed at $\theta = 70°$ and (a) $\varphi = -45°$ and (b) $\varphi = 0°$

the reflective sub-pixel is still a narrow-band circular polarizer with only one monochromatic quarter-wave film.

The corresponding polarization traces on the Poincaré sphere when viewed from $\theta = 70°$, $\varphi = -45°$ and $\theta = 70°$, $\varphi = 0°$ are shown in Figure 3.29. As expected, in both directions, the final polarization can be converted to overlap with the front polarizer's absorption axis at point **A** on the Poincaré sphere. In Figure 3.29(a) when viewed at $\theta = 70°$, $\varphi = -45°$, the first biaxial film moves the polarization from point **T** to point **A**. The following rear negative quarter-wave film rotates the polarization from point **A** to **B**, which is then fully converted back to point **A** by the front positive quarter-wave film. In Figure 3.29(b) when viewed at $\theta = 70°$, $\varphi = 0°$, the first rear biaxial retarder does not react to the light polarization change, as its n_x axis is set along the absorption axis of the rear linear polarizer. The two quarter-wave films cancel each other, taking the final polarization back to point **A**.

Here, compared with the conventional configuration using both positive uniaxial quarter-wave films, we can see the tremendous advantages of including a negative uniaxial film in the design. First, the optical axes of the positive and negative films are aligned parallel to each other. Hence, these two films will always be able to fully cancel each other with overlapped polarization traces and similar phase retardation values regardless of the incident azimuthal and polar angles. In contrast, when two quarter-wave films, both with positive birefringence, are set to be orthogonal (one at 45° and another at 135°), the related compensation is incident-angle-dependent. When viewed at $\theta = 70°$, $\varphi = -45°$, their optical axes are still perpendicular to each other, but the off-axis phase retardation values become different. More specifically, in the phase retardation equation of a uniaxial A-film $\Gamma_a = \frac{2\pi}{\lambda} d \left[n_e \sqrt{1 - \frac{\sin^2\theta_0 \sin^2\phi_n}{n_e^2} - \frac{\sin^2\theta_0 \cos^2\phi_n}{n_o^2}} - n_o \sqrt{1 - \frac{\sin^2\theta_0}{n_o^2}} \right]$, their relative values ϕ_n (relative angle between the optical axis and the azimuthal angle of the incident light) are different in this direction. On the other hand, when viewed at $\theta = 70°$, $\varphi = 0°$, the off-axis retardations are the same, but their optical axes are no longer orthogonal to the incident wave plane, making the polarization change traces diverge. The calculated angular light leakage of the transmissive sub-pixel from the above circular polarizers is shown in Figure 3.30 (the reflective sub-pixel is similar to that in Figure 3.26(b)). The maximum normalized light leakage of the transmissive sub-pixel is less than 1.8×10^{-3}, making this configuration quite attractive for wide-view transmissive or transflective LCDs, except for the narrow band property in the reflective sub-pixel.

In the above-mentioned circular polarizer design, the associated compensation mechanisms are quite direct and clear: positive and negative quarter-wave

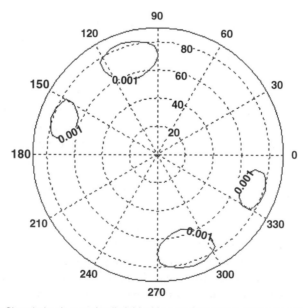

Figure 3.30 Simulated angular light leakage (normalized to maximum parallel polarizer transmittance) at $\lambda = 550$ nm for the transmissive sub-pixel

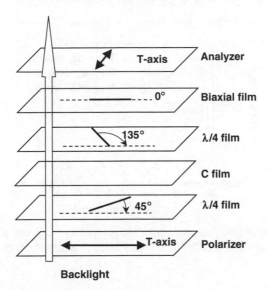

Figure 3.31 Circular polarizers using one biaxial film and a positive C-film for compensation

films compensate for each other and the additional biaxial film (or uniaxial films) compensates for the effective polarizer angle deviation. Yet, in another example, we will introduce a new configuration to obtain a wide viewing angle using different methodology, as shown in Figure 3.31 [26]. Compared with the simple circular polarizer configuration using a linear polarizer and a quarter-wave plate in Figure 3.23, this device includes one more positive C-film and one biaxial film. The positive C-film could be realized by the overall phase retardation from a negative C-film and the VA LC cell. The related compensation principles will be introduced later. By adjusting the film parameters, the spectral light leakage of the transmissive and reflective sub-pixels is shown in Figure 3.32. Similarly, the transmissive sub-pixel is well compensated to show negligible light leakage over a wide range of wavelengths. The reflective sub-pixel also shows a limited improvement in the off-axis spectral light leakage.

The detailed compensation mechanism is illustrated in Figure 3.33, where the corresponding polarization traces on the Poincaré sphere when viewed from $\theta = 70°$, $\varphi = 0°$ and $\theta = 70°$, $\varphi = -45°$ are plotted. When light is incident from $\theta = 70°$, $\varphi = 0°$, after traversing the rear linear polarizer, the backlight with a polarization state at point **T** will be first moved to point **B** by the rear positive quarter-wave film, then by a properly designed retardation value of the positive C-film, we can convert the light from point **B** to point **C**, from

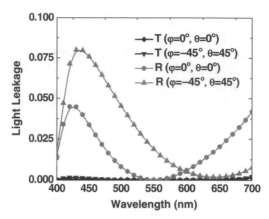

Figure 3.32 Calculated spectral light leakage of both transmissive and reflective sub-pixels (without normalization) in the circular polarizer configuration shown in Figure 3.31

which the following positive quarter-wave film can move the polarization state from point **C** to point **A**. Here, because the biaxial film has its n_x axis aligned in a direction parallel to the absorption direction of the front linear polarizer, it does not change the light polarization at point **A**; thus, a perfect

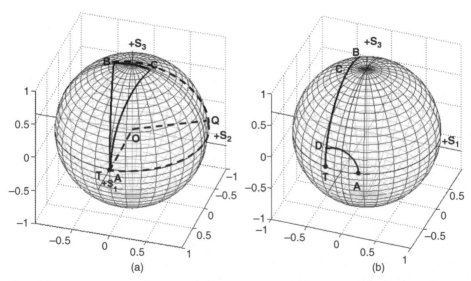

Figure 3.33 (a) Polarization state trace on the Poincaré sphere at $\theta = 70°$ when viewed at (a) $\varphi = 0°$ and (b) $\varphi = -45°$

dark state in this direction $\theta = 70°$, $\varphi = 0°$ can be obtained. The insertion of this positive C-film mainly functions to compensate the non-orthogonal optical axes of the two quarter-wave films (see Figure 3.25(b)).

In the other direction with $\theta = 70°$, $\varphi = -45°$ as shown in Figure 3.33(b), point **T** representing the rear linear polarizer transmission axis departs from point **A**, representing the absorption axis of the front linear polarizer. Light passing the rear linear polarizer will be first rotated from point **T** to point **B** after passing through the rear $\lambda/4$ film, and is then further moved to point **C** by the positive C-film. In sequence, the front $\lambda/4$ film converts it from point **C** back to point **D**. Here, the front biaxial film with proper film parameters functions to convert light from point **D** to point **A**, leading to perfect compensation in this direction as well.

From the polarization traces shown in Figure 3.33(a), we can analytically determine the phase retardation value of the positive C-film based on spherical trigonometry. The target is to obtain the spherical angle $\angle$**BAC** (or $\angle$**BTC**) or arc $\overline{BC}$ that is associated with the phase retardation of the C-film. On the Poincaré sphere, point **T** and point **A** overlap each other when $\varphi = 0°$. Point **Q** stands for the optical axis of the rear quarter-wave film, and the relative angle between the effective transmission axis of the rear linear polarizer and the optical axis of the rear quarter-wave film on the wave plane can be obtained by setting $\phi_1 = \phi_k = 0°$ and $\phi_2 = 45°$ in Equation (3.1)

as $\cos^{-1}\left(\frac{\sqrt{2}}{2}\frac{\sqrt{1-\sin^2\theta_0/n_p^2}}{\sqrt{1-\sin^2\theta_0/(2n_p^2)}}\right)$ (here the angle is on the wave plane in x-y-z

coordinates). With $\theta_0 = 70°$ and n_p of the polarizer at 1.5, $\angle$**TOQ** $= 2 \times 52.07° = 104.14°$ on the Poincaré sphere (a relation of x2). In the spherical triangle **QBT**, arc $\overline{QB}$ = arc $\overline{QT}$, and they are both equal to $\angle$**TOQ** (~ 1.817). The spherical angle $\angle$**TQB** is equal to the phase retardation of the quarter-wave film in the specific incident direction, which can be obtained as

$\Gamma_a = \frac{2\pi}{\lambda}d\left[n_e\sqrt{1-\frac{\sin^2\theta_0\sin^2\phi_n}{n_e^2}-\frac{\sin^2\theta_0\cos^2\phi_n}{n_o^2}}-n_o\sqrt{1-\frac{\sin^2\theta_0}{n_o^2}}\right]$, where ϕ_n is the align-

ment angle of the quarter-wave film that is equal to 45° with respect to the incident azimuthal angle at 0°. From this equation, we can obtain the spherical angle $\angle$**TQB** of about 92.70° with n_e and n_o of the positive uniaxial quarter-wave film at 1.5110 and 1.5095 at $\lambda = 550$ nm, respectively. Based on spherical trigonometry, we can obtain the arc length of $\overline{BT}$ from the following equation: $\overline{BT} = \cos^{-1}[\cos(\overline{QT})\cos(\overline{QB}) + \cos(\Gamma_a)\sin(\overline{QT})\sin(\overline{QB})]$ as ~ 1.5555. With three arc lengths in the triangle **QBT**, we can then calculate the spherical angle $\angle$**BTQ** to be around 104.36°. Because arc $\overline{BT}$ and arc $\overline{CT}$ are symmetrical to each other, we can obtain $\angle$**BAC** as $2\angle$**BTQ** $- \pi$ ($\sim 28.71°$), which is equal to the required phase retardation value of the negative C-film at $\Gamma_c = \frac{2\pi}{\lambda} d \left[n_o \sqrt{1 - \frac{\sin^2\theta_0}{n_e^2}} - n_o \sqrt{1 - \frac{\sin^2\theta_0}{n_o^2}} \right]$. Therefore, if the refractive indices n_e and n_o for the positive C-film are set at 1.5024 and 1.4925, the $d(n_e - n_o)/\lambda$ value of the C-film is about 0.158. Interestingly, we also found this value remains roughly constant even if the polar angle θ varies from 1° to 89° [26].

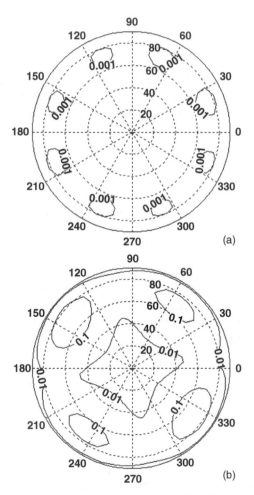

Figure 3.34 Angular-dependent light leakage of (a) the transmissive sub-pixel and (b) the reflective sub-pixel for the circular polarizer configuration shown in Figure 3.31

In the other incident direction where $\theta = 70°$ and $\varphi = -45°$, with fixed phase retardation values of the quarter-wave films and the positive C-film, the polarization state of incident light before it impinges on the front biaxial film is solely determined at the point **D** in Figure 3.33(b). Therefore, we can tune the N_z factor $\left(\frac{n_x - n_z}{n_x - n_y}\right)$ and the in-plane phase retardation $d(n_x - n_y)/\lambda$ value of the biaxial film to make the polarization coincide with the absorption direction at point **A**. From numerical calculations based on the extended Jones matrix and polarization trace plot on the Poincaré sphere, we find that when N_z is about 0.35 and $d(n_x - n_y)/\lambda$ is about 0.34, the compensation is best, showing minimal leakage.

We should point out that when the above circular polarizer is applied to MVA-based transmissive or transflective LCDs, the positive C-film can actually be realized by a partial compensation of the negative C-film to the VA LC cell, which behaves like a positive C-film. In other words, the overall phase retardation value $d(n_e - n_o)/\lambda$ of the VA LC cell and the negative C-film should be about 0.158 at $\lambda = 550$ nm. Figure 3.34 plots the angular light leakage from the transmissive and reflective sub-pixels of the above circular polarizer configuration. For the transmissive sub-pixel the maximum light leakage is $\sim 1.4 \times 10^{-3}$ after normalization to the maximum axial transmittance of two parallel linear polarizers. For the reflective sub-pixel, 10% normalized light leakage occurs at about 50°, which should provide a reasonably wide viewing angle for the reflective display as well. With this polarization configuration, a contrast ratio of 200 : 1 can be achieved up to angles greater than 50° in the polar direction for a transmissive MVA LCD [26].

3.4 Summary

In this chapter we have studied light polarization in an LCD system based on the Poincaré sphere. With visualized polarization trajectories, the associated optics and mechanisms of light passing through uniaxial and biaxial films become quite intuitive. We also studied several typical optical compensation schemes for wide-view linear polarizers and circular polarizers. Illustrating the polarization traces on the Poincaré sphere not only helps us to better understand the origins of off-axis light leakage from linear polarizers and circular polarizers, but also gives us indications how to optimize their performance. These device design concepts and their associated methodologies will undoubtedly have a significant impact on transflective LCD technologies, especially those based on multi-domain vertical alignment LC cells.

References

[1] Huard, S. (1997) *Polarization of Light*. John Wiley & Sons, Inc., New York.
[2] Bigelow, J.E. and Kashnow, R.A. (1977) Poincaré sphere analysis of liquid crystal optics. *Appl. Opt.*, **16**, 2090.
[3] Zhu, X., Ge, Z. and Wu, S.T. (2006) Analytical solutions for uniaxial-film-compensated wide-view liquid crystal displays. *J. Disp. Technol.*, **2**, 2–20.
[4] Yeh, P. and Gu, C. (1999) *Optics of Liquid Crystal Displays*. John Wiley & Sons, Inc., New York.
[5] Chen, J., Kim, K.-H., Jyu, J.-J., Souk, J.H., Kelly, J.R. and Bos, P.J. (1998) Optimum film compensation modes for TN and VA LCDs. *SID Dig. Tech. Papers*, **29**, 315–318.
[6] Anderson, J.E. and Bos, P.J. (2000) Methods and concerns of compensating in-plane switching liquid crystal displays. *Jpn. J. Appl. Phys.*, part 1, **39**, 6388–6392.
[7] Saitoh, Y., Kimura, S., Kusafuka, K. and Shimizu, H. (1998) Optimum film compensation of viewing angle of contrast in in-plane-switching-mode liquid crystal display. *Jpn. J. Appl. Phys.*, part 1, **37**, 4822–4828.
[8] Khoo, I.-C. and Wu, S.T. (1993) *Optics and Nonlinear Optics of Liquid Crystals*, World Scientific, Singapore.
[9] Li, J. and Wu, S.T. (2004) Extended Cauchy equations for the refractive indices of liquid crystals. *J. Appl. Phys.*, **95**, 896.
[10] Fujimura, Y., Kamijo, T. and Yoshimi, H. (2003) Improvement of optical films for high performance LCDs. *Proceedings of SPIE*, **5003**, 96–105.
[11] Yang, Y.-C. and Yang, D.K. (2008) Achromatic reduction of off-axis light leakage in LCDs by self-compensated phase retardation (SPR) film. *SID Tech. Digest*, **39**, 1955–1958.
[12] Ishinabe, T., Miyashita, T. and Uchida, T. (2002) Wide-viewing-angle polarizer with a large wavelength range. *Jpn. J. Appl. Phys.*, part 1, **41**, 4553–4558.
[13] Yoshida, H., Tasaka, Y., Tanaka, Y., Sukenori, H., Koike, Y. and Okamoto, K. (2004) MVA LCD for Notebook or Mobile PCs with High Transmittance, High Contrast Ratio, and Wide Angle Viewing. *SID Tech. Digest*, **35**, 6–9.
[14] Yang, Y.-C., Choi, J.Y., Kim, J., Han, M., Chang, J., Bae, J., Park, D.-J., Kim, S.I., Roh, N.-S., Kim, Y.-J., Hong, M. and Chung, K. (2006) Single Cell Gap Transflective Mode for Vertically Aligned Negative Nematic Liquid Crystals. *SID Tech. Digest*, **37**, 829–831.
[15] Ge, Z., Zhu, X., Lu, R., Wu, T.X. and Wu, S.T. (2007) Transflective liquid crystal display using commonly biased reflectors. *Appl. Phys. Lett.*, **90**, 221111.
[16] Lu, R., Ge, Z. and Wu, S.T. (2008) Wide-view and single cell gap transflective liquid crystal display using slit-induced multidomain structures. *Appl. Phys. Lett.*, **92**, 191102.
[17] Lin, C.H., Chen, Y.R., Hsu, S.C., Chen, C.Y., Chang, C.M. and Lien, A. (2008) A novel advanced wide-view transflective display. *J. Disp. Technol.*, **4**, 123.

[18] Pancharatnam, S. (1956) Achromatic combinations of birefringent plates. *Proc. Ind. Acad. Sci. A*, **41**, 130–144.

[19] Ishinabe, T., Miyashita, T. and Uchida, T. (2001) Design of a quarter wave plate with wide viewing angle and wide wavelength range for high quality reflective LCDs. *SID Tech. Digest*, **32**, 906–909.

[20] Yoshimi, H., Yano, S. and Fujimura, Y. (2002) Optical Films for Reflective LCDs to Achieve High Image Quality. *SID Tech. Digest*, **33**, 862–865.

[21] Hong, Q., Wu, T.X., Zhu, X., Lu, R. and Wu, S.T. (2005) Designs of wide-view and broadband circular polarizers. *Opt. Express*, **13**, 8318.

[22] Hong, Q., Wu, T.X., Lu, R. and Wu, S.T. (2005) Wide-view circular polarizer consisting of a linear polarizer and two biaxial films. *Opt. Express*, **13**, 10777.

[23] Lin, C.H. (2006) Optical compensation of a high transmittance MVA-LCD. *SID Tech. Digest*, **37**, 1075–1078.

[24] Lin, C.H. (2007) Extraordinarily wide-view and high-transmittance vertically aligned liquid crystal displays. *Appl. Phys. Lett.*, **90**, 151112.

[25] Ge, Z., Jiao, M., Lu, R., Wu, T. X., Wu, S. T., Li, W. Y., and Wei, C. K. (2008) Wide-view and broadband circular polarizers for transflective liquid crystal displays. *J. Disp. Technol.*, **4**, 129.

[26] Ge, Z., Lu, R., Wu, T.X., Wu, S.T., Lin, L., Hsu, N.C., Li, Y. and Wei, C.K. (2008) Extraordinary wide-view circular polarizers for liquid crystal displays. *Opt. Express*, **16**, 3120.

[27] Lin, C.H. (2008) Optically compensated circular polarizers for liquid crystal displays. *Opt. Express*, **16**, 13276.

4

Wide-view Transflective LCDs

4.1 Overview

Presently, the majority of mobile LCDs available on the market use transmissive twisted nematic (TN) mode because of its low cost, high brightness, simple manufacturing process, and high yield. Most transflective mobile displays still use dual-cell-gap ECB (electric-field-controlled birefringence) LCDs or just TN cells with a transflective sheet laminated on to the rear polarizer. However, because of their inherent LC orientation characteristics, the image quality of uncompensated TN or ECB LCDs exhibits a relatively low contrast ratio, narrow viewing angle, and grayscale inversion. Thus, TN or ECB-based mobile displays are no longer adequate to meet the increasing demands of customers. Instead, new mobile displays based on wide-view multi-domain vertical alignment (MVA) and in-plane switching (IPS) or fringe-field switching (FFS) modes are emerging and will begin to dominate the high-end market in the near future. Hence, in this chapter, we will introduce and discuss in detail mobile LCDs based on these two major wide-view technologies for both transmissive and transflective applications.

Transflective Liquid Crystal Displays   Zhibing Ge and Shin-Tson Wu
© 2010 John Wiley & Sons, Ltd

4.2 Transflective LCD Using MVA Mode

4.2.1 MVA Technology Overview

MVA LCD technology exhibits unique advantages over all other modes (TN, IPS, or pi-cell) in terms of its inherent high contrast ratio and rubbing-free fabrication process, making it the dominant technology for TV applications. Its transmittance is continually improving, because of high aperture designs, polymer-sustained alignment and the use of circular polarizers, generating ever-increasing interest for mobile displays. In this section, the cell design characteristics and considerations for large-panel and small-panel MVA LCDs will be addressed.

A single-domain VA cell is asymmetrical when viewed from different directions, as illustrated in Figure 4.1. Severe grayscale inversion occurs owing to the asymmetry of the LC director profile and optical anisotropy of the LCs. For example, for a viewer on the right-hand side, looking into the single-domain VA from a certain polar direction, a darker image would be observed from the voltage-on state than that at $V = 0$. But this asymmetry could be eliminated by forming multi-domain structures, as shown in the right-hand plot. Here, viewers on both the left and right sides see a similar averaged LC director profile, and thus nearly identical brightness. To remove the azimuthal angle dependence, a four-domain structure, each with a single domain aligned at $\pm 45°$ from the linear polarizer's transmission axis, can be used. In some applications such as TVs, eight-domain alignment has been implemented in order to reduce the off-axis gamma curve shift. The eight-domain structure is actually comprised of two four sub-domains that are generated by different voltages to average out the tilt angles (or polar angles).

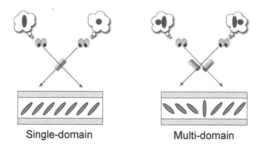

Single-domain   Multi-domain

Figure 4.1 Illustration of different viewing angle properties for single-domain and multi-domain VA cells

To generate multi-domains in a VA LC cell, the most direct method is to apply surface rubbings in different directions. But it is not practically possible to accomplish this in each sub-pixel (even for TVs with a sub-pixel size of about $100\,\mu m \times 300\,\mu m$), thus this method has not been adopted in mass production. Instead, the most commonly adopted method for generating multi-domains is to form patterned structures on the electrodes. The first commercially successful MVA LCD technology was developed by Fujitsu [1] and involved forming multi-domain LC profiles by using protrusions on both substrates (MVA phase I, as shown in Figure 4.2(a)). With a small inclination initiated by the protrusion surface, the LC molecules there have a preferred direction in which to reorient once voltage is applied and multi-domains are automatically formed without rubbing. The drawback is the sophisticated process for forming protrusions on both substrates, which could also lower the contrast ratio and the useful aperture of the display. With continuous research efforts on this technology, Fujitsu later proposed an improved version of the MVA technology by replacing protrusions on the rear TFT substrate with slits (MVA phase II, as shown in Figure 4.2(b)) [2]. In this improved design, protrusions are only processed on the front color filter substrate, which greatly reduces the fabrication complexity and increases the contrast ratio. When a relatively high voltage is applied, fringing electric fields are generated in the vicinity of the slits, driving the LC molecules to tilt away from the slits. The tilt from the slits along with the adjusting effect from the protrusion surfaces automatically forms multi-domains. Lien *et al.* first proposed to use slits on both substrates to guide the LC reorientation [3–5], and Samsung later commercially developed the patterned vertical alignment (PVA) technology using slits only [6]. In PVA structures, the adjustment of LC reorientations comes from the fringing electric field near the slits. Without any opaque protrusions, the LC molecules are perpendicular to the substrates in the dark state, thus the contrast ratio of PVA is extremely high.

The response of LC molecules in an MVA cell to external electric fields (taking Fujitsu's design in Figure 4.2(b) as an example) is illustrated in

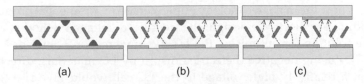

(a)   (b)   (c)

Figure 4.2 Cross-sectional configuration of (a) an MVA cell with protrusions, (b) an MVA cell with protrusions and slits, and (c) a PVA cell

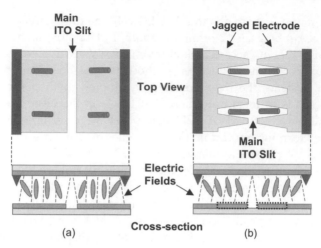

Figure 4.3 Top view and cross-sectional view of (a) a conventional MVA cell, and (b) an MVA cell with jagged electrodes

Figure 4.3(a). When a voltage is applied between the pixel and common electrodes, the LCs near the protrusions (because of the initial pre-tilt on the protrusion surfaces) and slits (because of the fringing electric fields there) tilt down quickly, but LCs in between stay unperturbed initially, as electric fields there are normal to the substrate surfaces and also parallel to the LC alignment. Gradually, the tilting of the LC molecules propagates horizontally from the protrusions and slits towards the central region to perturb other LCs. This propagation process leads to the slow response time observed in MVA LCDs. To increase the switching speed, Fujitsu later proposed an MVA-premium design with jagged pixel electrodes, as shown in Figure 4.3(b) [7] to generate simultaneous reorientation. The jagged electrode generates fringing fields deeper into the central part of the display cell, and more LCs can be rotated by the electric fields at a similar pace, unlike the previous slow propagation process. As a further extension, the major slits in a PVA cell (Figure 4.2(c)) can be replaced by jagged shapes on both substrates to improve the response time, and this is now the major technology implemented in commercial products [8].

In addition to electrode design, the overdrive voltage method is another effective solution for optimising the LC dynamic response. But for MVA or PVA LCDs using negative dielectric anisotropy LC materials, direct application of a relatively high voltage to the LC cell would give rise to some LC disclinations, causing undesired electro-optical behavior, which would lengthen the response time. Thus, a special driving scheme needs to be

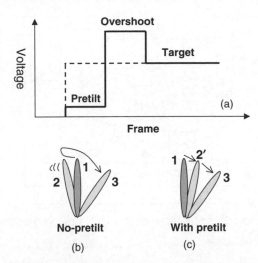

Figure 4.4 Illustration of (a) driving schemes with and without pre-tilt and overshoot voltages, (b) LC director response under regular driving, and (c) LC director response with pre-tilt and overshoot (redrawn from (9))

designed to account for the unique dynamic properties of PVA. Figure 4.4(a) illustrates the different driving schemes of a regular PVA LCD, where the dashed line represents the regular driving signal and the solid line stands for the new method with a pre-tilt signal and an overshoot one [9]. Figures 4.4(b) and 4.4(c) depict the LC director responses from these two schemes, respectively. In a regular PVA cell, LC directors are all initially aligned at position 1; that is, normal to the substrate. When a relatively high voltage is suddenly applied in the regular driving method, without a desired reorientation direction, some LC directors in the central electrode region between two adjacent slits are randomly pushed to position 2; that is, opposite to the final designated position 3. The realignment from position 2 to position 3 involves both polar and azimuthal reorientations at a slow speed.

One method is to apply a small bias voltage on all the pixels in the dark state to give the LCs there a pre-orientation, independent of the next image state in the following frames. But even this low bias voltage could cause the inherent high contrast of the VA mode to deteriorate. To solve this issue, Samsung proposed a new driving waveform, as shown in Figure 4.4(a) [9]. In this new method, the typical black voltage at gray level 0 is applied to the dark pixels to ensure high contrast. When the gray level of a dark pixel in the next frame is changed, an initial pre-tilt signal starts to be applied to that dark pixel to prepare the LCs there. An overshoot voltage is also applied in the coming

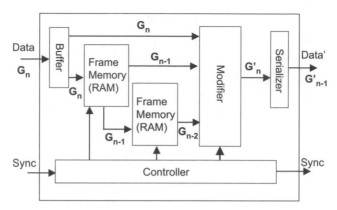

Figure 4.5 Basic logic control flow of the DCII driving method for PVA mode (redrawn from (9))

frame to speed up the LC director change from the pre-tilt state to the final target gray level profile without causing improper LC reorientations. The logic control flow is illustrated in Figure 4.5. First, the image G_{n-1} is compared with the previous image G_{n-2} and the next one G_n. If an addressed pixel is black in G_{n-1} and bright in G_n, then the modifier changes the voltage for that pixel at frame $n-1$ to a low voltage (this is the pre-tilt signal shown in Figure 4.4(a)). Additionally, in the next frame (the previous G_{n-1} is now indexed as G_{n-2}), by comparing the new G_{n-1} and G_{n-2}, the overshoot signal on this single pixel can be determined and the voltage for G_{n-1} can be changed by the modifier. From this method, we find that a small pre-tilt voltage is only applied for one frame onto those individually targeted pixels, and it will not affect other black pixels that will stay dark for a longer time. Hence, the pre-tilt voltage gives LCs a desired pre-orientation from position 1 to position 2′, where that special pixel still appears dark and is ready for a bright state; and the following overshoot voltage quickly switches the LCs from position 2 to the final position 3, as illustrated in Figure 4.4(c). With this method, fully dark to fully bright switching can be reduced to less than 8 ms for a real PVA panel, while the regular driving method needs more than 25 ms [9]. In addition, a more uniform gray-to-gray response time can be obtained.

Besides the pre-tilt-driving method discussed above, the polymer-sustained surface alignment (PSA) method has recently emerged as a means of inducing initial LC pre-tilt throughout MVA and PVA cells [10–16]. This method was first reported by Hanaoka *et al.* [10] from Fujitsu and was later widely incorporated into panel production by LCD manufacturers. The process flow, which usually consists of two to three major steps, is shown

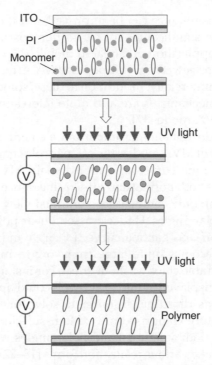

Figure 4.6 PSA process flow in an MVA cell (redrawn from (10))

in Figure 4.6. First, a host VA LC mixture is doped with a photosensitive monomer at a very low concentration (usually less than 0.2 wt%). Then a small bias voltage (slightly above threshold) is applied to the LC cell to make the LCs near the surfaces incline at small pre-tilt angles by the fringing electric fields near the ITO slits on the substrate. To cure the monomers, the cell is simultaneously exposed to UV light. During this curing process, the monomers will gradually migrate from the bulk toward the front LC alignment layer surface, from where UV light impinges. The cured polymer generates a small pre-tilt angle within each domain to adjust the LC orientation. Finally, the bias voltage is removed and UV exposure continues for a while to further remove the residual monomers in the bulk region. Most importantly, with all the LCs having a small but well-defined pre-tilt direction within each sub-domain for reorientation, they can now respond simultaneously to the applied electric fields and exhibit a much faster switching speed. In addition, with this method, protrusions can be removed and only slits on one substrate (not shown in the plot) are necessary to form multi-domains in the voltage-on state. Thus, the transmittance can be improved and the fabrication process greatly simplified to significantly cut fabrication costs. Recently, this technology has shown application potential in mobile displays [12, 13].

We have briefly reviewed several popular MVA technologies that have been introduced commercially. In addition to these, some other MVA designs and their guiding mechanisms are also quite interesting and enlightening, such as the biased VA mode [17].

Below, we will address another important element affecting the electro-optical performance of MVA modes: the off-axis color gamma shift. Figure 4.7 shows the color gamma curves of a four-domain MVA under linear and circular polarizers at both normal and off-axis incidence. In principle, a four-domain structure can only compensate azimuthal viewing directions, but is not sufficient for polar angles. Under crossed linear polarizers, the discrepancy between the off-axis gamma curve (LP$-60°$) and the normal gamma curve (at $\gamma = 2.2$) becomes quite large from low to mid gray levels. This generates an undesirable color washout effect. For instance, a natural human face that is usually displayed at mid gray levels could appear 'whitish' when viewed at an off-axis direction. In order to solve this problem, an eight-domain MVA structure comprised of two four-sub-domains with different voltages to average both azimuthal and polar angles would be an effective method; this could take various realization forms [18–22]. In addition, mobile

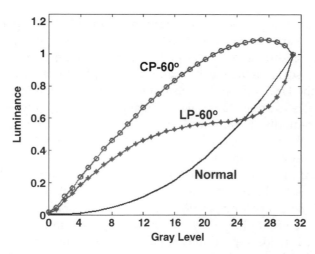

Figure 4.7 Color gamma curves of a four-domain MVA under linear and circular polarizers

MVA LCDs usually use circular polarizers to enhance transmittance and block internal reflections from ambient light, making the color gamma shift even worse than in displays using linear polarizers.

In Figure 4.7, the off-axis gamma curve (CP−60°) shows a much larger discrepancy than that obtained at normal incidence, owing to the approximately circular polarization of the incident light. This would be a big disadvantage for mobile MVA LCDs (both transmissive and transflective types) when using circular polarizers. Utilizing two four-sub-domains helps to reduce the gamma shift for mobile MVA LCDs [23], but the improvement is rather limited and it also dramatically increases the fabrication difficulty, especially for high-resolution devices.

4.2.2 Mobile MVA Technology

For mobile LCD applications, the pixel size is typically quite small (50 μm × 150 μm or smaller); thus, the electrode pattern used to introduce multi-domains is quite different from those for large-screen TVs. Here, we will first look at typical pixel designs for mobile MVAs and then discuss their design considerations and related electro-optical properties.

Figure 4.8 is a plot of three different types of mobile MVA (or PVA) pixel layout. In mobile displays, the pixel size is limited, so to improve aperture ratio, the multi-domain structure is formed by centered protrusions or ITO openings, instead of the zigzag-shaped electrode slits used in large-panel displays. Forming a round protrusion on the front color filter (CF) substrate, as shown in Figure 4.8(a), is a very effective method of guiding LC molecules to tilt uniformly in the azimuthal direction. Since adjacent pixels might have different voltage polarities, to avoid the influence of fringe fields from adjacent pixels, protrusions need to be processed on the front substrate,

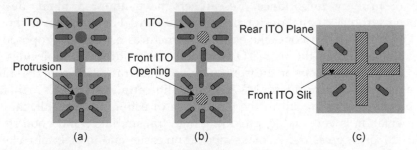

Figure 4.8 Pixel layout for mobile MVA (or PVA) technologies using (a) protrusions, (b) circular ITO openings, and (c) crossed ITO openings on the front color filter substrate

rather than on the rear one. Otherwise, the reorientation direction of LC molecules in the present pixel adjusted by the rear protrusion surfaces will be opposite to the tendencies induced by the fringe fields near the edges of the present pixel and those adjacent to it.

Another method is to create ITO openings on the front substrate, as shown in Figure 4.8(b). In addition to protrusions or ITO openings, in some designs patterned openings are also formed on the rear ITO electrode to better adjust the LC director orientation [24, 25].

Comparing the methods using protrusions and openings, protrusion structures generate more reliable LC director profiles when voltages are applied and their inherent response times are faster due to the strong adjusting effect from the protrusion surfaces. For structures using ITO openings on the front CF substrate, in order to provide better touch restoring capability, the radius of the opening is quite critical and a larger value is better [25], but the transmittance may be reduced.

To improve the surface transmittance, circular polarizers are typically employed in such mobile MVA LCDs. As a result, the color washout problem becomes severe (as illustrated in Figure 4.7), as the off-axis gamma curve deviates significantly from the normal one. To suppress color washout, two sub-pixels with different voltages could be used to create different tilt angles, but this is rather complicated for small pixels. As an alternative, researchers propose to design a special mobile MVA structure that can use a linear polarizer. Figure 4.8(c) shows the pixel layout of the design with a cross-shaped ITO opening on the front substrate [26]. Thus, the LC directors will mainly tilt as shown in the plot with good transmittance under crossed linear polarizers. This way, the color washout problem can be minimized to acceptable levels for mobile MVA LCDs.

For mobile MVA technology, LC on-state stability and high transmittance are of primary importance. Researchers have adopted many design concepts from TV panel design in mobile displays. For example, as shown in Figure 4.9(a), fish-bone-like electrode structures have been proposed to create fine slits on the rear ITO electrodes for guiding LC molecules, in addition to protrusions on the front color filter substrate [27]. From the study of jagged pixel electrodes for MVA [7], fine slits can adjust the LCs to tilt in a stable way along the slit in the lengthwise direction. As a result, the LC reorientation is very stable, providing high transmittance and good resistance to touch pressure. To make fabrication easier and to maintain a large aperture, the front substrate protrusions are removed and a polymer-sustained surface alignment method is also used in small-panel mobile LCDs, as shown in Figure 4.9(b) [12, 16]. Fine slits are formed only on the rear

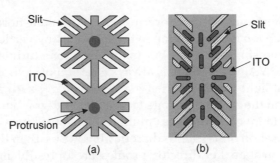

Figure 4.9 Pixel layout for advanced mobile MVA technologies using (a) protrusions on the front substrate and fine slits on the rear substrate, and (b) slits only on the rear substrate but with polymer-sustained surface alignment

ITO pixel electrode and the front ITO common electrode is planar in shape. The fringe fields near the slits on the rear substrate guide LC molecules to form multiple domains, which are then anchored by surface polymerization. This seems a promising technology for mobile MVA LCDs, which could provide both high transmittance and strong restoring forces after external touch pressure is released.

For these mobile MVA (or PVA) technologies, to include the transflective feature for outdoor sunlight readability, a reflective region is also formed in each pixel. To keep the aperture of the transmissive sub-pixel as large as possible, the reflective sub-pixel is usually formed on the storage capacitor region. To match the gamma curves of the transmissive mode and the reflective mode, we could use the double-cell-gap method, the capacitor-shielding-voltage method, or some other approaches as discussed in Chapter 2 [28–32].

4.3 Transflective LCD Using IPS Mode

4.3.1 IPS and FFS Technology Overview

In-plane switching (IPS) [33–35] and fringe-field switching (FFS) [36] modes use horizontal electric fields to drive the homogeneously aligned LC molecules. Initially, the LC rubbing direction is set either parallel or perpendicular to one of the linear polarizer transmission axes, thus its dark state is independent of wavelength and temperature. At oblique incidence, the off-axis light leakage mainly originates from the deviated effective angle of the crossed polarizers, which could be compensated by uniaxial films or biaxial films, as described in Chapter 3. When a relatively high voltage is applied between the pixel and common electrodes, LC molecules are perturbed by horizontal electric (or fringe) fields to rotate horizontally, thus gray-level inversion is weaker than in the modes using vertical fields to tilt the LCs like the single-domain VA mode. In a longitudinal field-driven LCD mode like VA or TN, there is always a certain off-axis direction from which incident light could experience a low or nearly zero phase retardation, even when the applied voltage targets the mid-gray level. This could create grayscale inversions unless multi-domain structures or inhomogeneous compensation films such as discotic films are used. In IPS or FFS cells, typically the on-state LC director profiles are optically modeled as an effective uniaxial retardation film with its effective optical axis at ~45° from its initial alignment axis. But this is not an accurate description and it could obscure the origins of some unique electro-optical properties of the IPS or FFS mode, such as weak color dispersion. In reality, due to the strong surface anchoring on both substrates, the LC profile is like two TN cells with opposite twist senses. From rear to front substrate, the LC directors deviate gradually away from the initial rubbing angle to a maximum rotational angle, then they gradually rotate back to the initial rubbing direction at the front substrate surface. As discussed in Chapter 1, the maximum rotational angle, which critically determines the maximum light transmittance, varies with driving voltage and horizontal position. In addition, the twist-like structures make IPS and FFS modes quite insensitive to wavelength dispersion. Owing to the inherent LC structures and working mechanism, IPS and FFS modes exhibit many attractive properties for mobile applications, such as wide viewing angle, weak gray-level inversion, weak gamma curve shift, and relatively uniform gray-to-gray response time. Following on from the basic working principle of IPS/FFS mode addressed in Chapter 1, we will study below the related technical aspects of IPS/FFS mode, such as aperture ratio enhancement by novel cell and electrode layout, and color washout reduction. After learning the basic physics of IPS/FFS LCDs, we will move on to look at sunlight-readable transflective LCDs employing IPS or FFS cells.

The conventional IPS cell configuration is shown in Figure 4.10, where interdigitated pixel and common electrode strips are formed on the same rear substrate to generate horizontal electric fields. In an early design, the gate line and common electrode were fabricated on the same layer, and the data line and pixel electrode strips were in a parallel plane in front of the common electrode. An alignment layer such as a polyimide layer is coated on the pixel electrode and is rubbed at an angle φ with respect to the lengthwise direction of the electrode strips. To reduce cross-talk between the data line and the driving electrodes, such as the pixel and common electrodes,

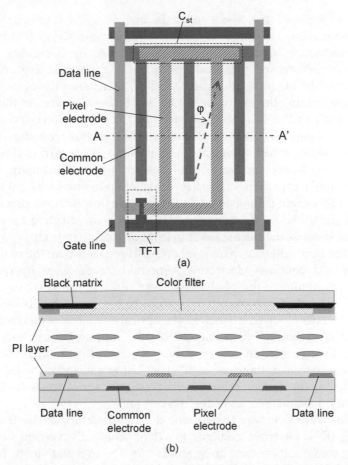

Figure 4.10 (a) Top view and (b) cross-sectional view of a conventional IPS cell

the data lines are usually formed away from the driving electrodes. In addition, a large area of black matrix is fabricated on the front color filter substrate to cover the region with data lines in order to obtain a high contrast ratio. These cause a significant loss in the effective aperture of the display. Meanwhile, to preserve the aperture ratio of the IPS LCD, an overlapping area of pixel and common electrodes is formed to work as a storage capacitor, as shown in Figure 4.10. In another approach, the storage capacitor is generated by overlapping the pixel electrode of one pixel and the gate line of the adjacent pixel in the next row.

From analysis of the above pixel layout of a conventional IPS cell, improvements in transmittance could mainly come from: (i) improving the transmittance in the existing open aperture by designing a novel electrode structure and using an optimized LC material (e.g., $\Delta\varepsilon < 0$) to better rotate the LC molecules there, and (ii) increasing the open aperture ratio by optimizing the electrode layout to decrease the area of the opaque regions, such as the storage capacitor and the black matrix on the front substrate. All new layouts must also minimize crosstalk from the data lines. However the major technological development with regard to IPS LCDs is closely related to aperture ratio and transmittance enhancement.

Several major approaches have been developed to improve light transmittance in the existing open aperture. Replacing the opaque pixel and/or common electrodes with transparent ITO ones is an effective approach. In an IPS cell, because the electric fields in front of the electrode strips are mainly vertical, the transmittance there is inherently low. Hence, in the early stages, the pixel and common electrodes were both made from opaque metal materials to minimize the photomask steps and cut costs. By using transparent ITO electrodes, transmittance could be improved by about 25% in a typical IPS cell [37, 38]. Nevertheless, the optical efficiency of an IPS cell using ITO electrodes is about 75% of that of a TN cell. Thus, the issue of how to create better LC director rotation in front of the electrodes plays an important role in enhancing the transmittance of an IPS cell. Using a negative $\Delta\varepsilon$ LC material could help, but the tradeoff is increased driving voltage owing to the electrode configuration characteristics of the IPS cell [39].

The most effective way to reorient the LC directors is the fringe-field switching (FFS) electrode configuration [36]. The early version of the FFS electrode configuration was quite similar to IPS, except that, to introduce substantial fringe fields, the electrode width and electrode spacing between adjacent patterned pixel and common electrode strips was made less than the LC cell gap. Later, a planar common electrode was developed to further improve the aperture ratio of the display by incorporating the capacitance between the patterned pixel electrodes and planar common electrode into the cell storage capacitor [40] (such an FFS cell configuration can be found in Chapter 1). In principle, the driving mechanisms of the conventional IPS cell and the FFS cell are quite different. In an IPS cell, the electrode spacing is usually much larger than the LC cell gap. Substantially horizontal fields are generated mainly in the region between electrode strips to rotate the LC molecules horizontally, and vertical fields in front of the electrode strips tend to tilt the LCs there vertically. In an FFS cell, the vertical distance (a few hundred nanometers) between the patterned pixel electrodes and the planar common electrode is much less than the cell gap, and the spacing distance between adjacent pixel electrodes is also less than or close to the LC cell gap. Thus, extremely strong horizontal fields are generated near the patterned pixel electrode edges to rotate the LCs there. With a small electrode dimension, the rotation of LCs near the electrode edges could lead to LC reorientation in other regions such as the symmetrical centers in front of the electrode strips and in front of the region between electrode strips, where vertical fields dominate [39]. As a result, the light efficiency of the FFS mode could reach over 95% of that of a TN cell when using a $-\Delta\varepsilon$ LC material, or over 85% if a $+\Delta\varepsilon$ LC were employed. Nowadays, the FFS electrode configuration is widely used in the IPS family as one of the major horizontal switching technologies for LCDs. But a tradeoff of the FFS cell is that it needs more fabrication masks than a typical IPS cell; thus the cost is higher.

Another way to improve transmittance is to increase the aperture ratio by reducing the opaque storage capacitor area (without sacrificing the C_{st} value) and the black matrix area. As mentioned earlier, in an FFS structure, part of the storage capacitor can be obtained from the capacitance of the overlapping region between the pixel and common electrodes. Thus, the remaining storage capacitor region using an opaque metal plate can be reduced to increase the aperture ratio. Because of the transmittance improvement from decreasing the black matrix covering the data line, the coupling between data line and adjacent pixel or common electrodes should remain minimal, so as not to sacrifice the contrast ratio of the display. Figure 4.11 shows two designs that could improve the aperture ratio of the IPS mode compared with that in Figure 4.10(b). In the conventional configuration in Figure 4.10(b), the data line is placed away from the neighboring pixel and common electrode strips to reduce the coupling capacitance. However, when one pixel is in the dark state and the data line for another pixel in the same column is at a relatively high voltage, fringe fields between the data line carrying a relatively high voltage for other pixels and the driving electrodes (targeted at zero voltage) in the dark pixel could still perturb the LCs locally, resulting in light leakage that requires a large black matrix to block. In contrast, in Figure 4.11(a), the data line is formed on a separate layer from the pixel and common electrodes with a thick passivation layer in between, and a wide common electrode is placed in front of the data line to shield the data-line-induced electric fields [38]. Thus, the coupling capacitance is reduced by the passivation layer, the influence of the dark state from the data line can be decreased, and the black matrix width can be reduced to improve the aperture ratio. Nevertheless, the common electrode covering the data lines needs to be sufficiently wide to shield the data line influence, causing a limited improvement in aperture ratio. In addition, a thick passivation layer could generate some fabrication problems.

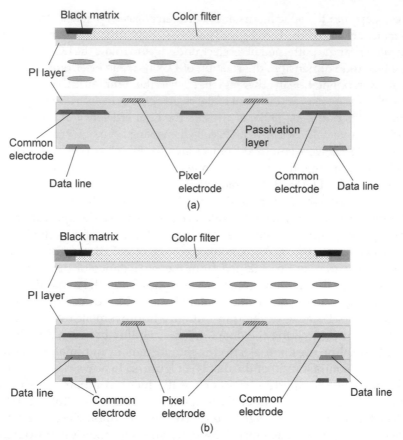

Figure 4.11 Cross-sectional view of a high aperture ratio IPS structure with (a) a thick passivation layer and a wide common electrode and (b) a passivation layer and additional common electrodes

In the improved design shown in Figure 4.11(b), the width of the shielding common electrode is further reduced, with the attraction of fields from two small common electrode strips behind the data line [41, 42]. Thus, the fields between the data line and in front of the common electrode strip into the LC region will be reduced. A similar concept of hiding or shielding the data line behind the common electrodes could also be applied to FFS LCDs to improve the aperture ratio of the display. In comparison to the TN cell, the FFS cell shows less transmittance in the open aperture region. However, its effective aperture ratio is higher than the TN cell, where a large opaque storage capacitor is needed. As a result, the overall transmittance in an FFS cell is higher.

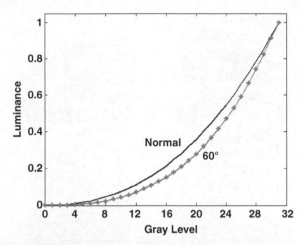

Figure 4.12 Color gamma curve performance of an FFS cell using $+\Delta\varepsilon$ LC material under linear polarizers

Another unique and attractive characteristic of IPS/FFS LCDs is their excellent color gamma performance compared with wide-view MVA technology. Figure 4.12 plots the color gamma curves at normal and 60° incidence, where the discrepancy is quite small. The result is obtained from an FFS cell with rectangular strip electrodes like that in Figure 4.13(a). Actually, for the FFS cell, the on-state LC directors, especially when a $+\Delta\varepsilon$ LC material is used, exhibit a multi-domain-like profile even in the single-domain electrode configuration owing to the complex fringe field patterns. However, for a single-domain FFS cell, a weak color shift can still be observed (a blue shift when viewed along the direction parallel to the on-state LC directors and a yellow shift when viewed along the direction perpendicular to the on-state LC directors). To minimize the color shift and widen the viewing angle of the FFS LCD, a dual-domain electrode configuration, as shown in Figure 4.13(b), could be considered. Here, single-domain and dual-domain mainly refer to the electrode profile. Besides using a dual-domain FFS configuration, where light efficiency may be reduced, an irregular electrode configuration could also be adopted in 'single-domain' FFS cells to generate more sub-domains to suppress the color shift [43].

For IPS and FFS cells, if the LC has $\Delta n \sim 0.1$ then the cell gap should be $\sim 3.5\,\mu m$ to $4.0\,\mu m$, and the response time will be over 20 ms owing to the small K_{22} associated with the in-plane rotation. Reducing the cell gap is the most straightforward way to reduce response time. However, there is a limit. If the cell gap is too thin, it is difficult to obtain good uniformity for large panels.

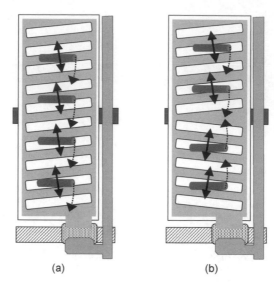

(a)   (b)

Figure 4.13 Schematic diagram of (a) a single-domain FFS LCD and (b) a dual-domain FFS LCD

When a relatively high voltage is applied, LC molecules are reoriented to form a dual-TN structure and a bright state is achieved. Figure 4.14 depicts the on-state azimuthal angle distributions at different horizontal positions in 2-μm and 4-μm FFS cells. The maximum rotation angle away from the initial rubbing direction varies at different horizontal positions, but is critical for the brightness, as discussed in Chapter 1. Corresponding to the electric field profile, the LC rotation near the electrode edge at position **C** is strongest. Obviously, even in the same position, a thicker cell has a larger reorientation angle and outputs a higher transmittance. This difference results from the surface anchoring effect in cells with different cell thicknesses, such that the LC rotation in a thinner cell is influenced more by the surface anchoring forces. In other words, reducing the cell gap will lead to decreased transmittance owing to the influence of the surface anchoring forces.

The response time, operating voltage, and maximum transmittance of FFS cells using different cell gaps are summarized in Table 4.1. As a reference, the transmittance of FFS cells is normalized to the value from two bare linear polarizers aligned parallel to each other. The TFT LC material employed has elastic constants $K_{11} = 9.7$ pN, $K_{22} = 5.2$ pN, and $K_{33} = 13.3$ pN, dielectric anisotropy $\Delta\varepsilon = +8.2$, birefringence $\Delta n = 0.1$ at $\lambda = 550$ nm, and rotational viscosity $\gamma_1 = 84$ mPa·s. Here, for the 2-μm cells, the birefringence is assumed to be 0.2 while other LC parameters remain unchanged. The surface rubbing

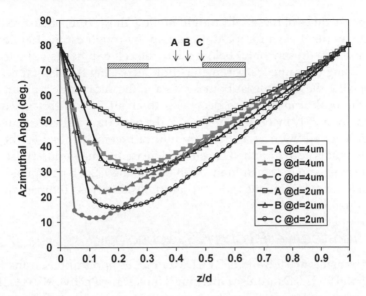

Figure 4.14 Azimuthal angle distributions at different positions in 2-μm and 4-μm FFS cells

angle of the LC layer is set at 80° with a pre-tilt angle of 2°, and strong anchoring is assumed on both surfaces. In accordance with the azimuthal angle plot, the thin FFS cells exhibit a lower transmittance. A finer electrode dimension (reducing the electrode width w and gap g) could effectively enhance the transmittance, but the tradeoff is increased driving voltage. The combination of $w = 2\,\mu m$ and $g = 3\,\mu m$ is attainable practically with present fabrication resolution capability. With these dimensions, the 2-μm FFS cell could output good normalized light efficiency (to the maximum possible value from two parallel linear polarizers) of about 86%. In addition,

Table 4.1 Calculated electro-optical properties of FFS cells using different cell gaps and electrode dimensions

	$w = 1\,\mu m, g = 1.5\,\mu m$		$w = 2\,\mu m, g = 3\,\mu m$		$w = 3\,\mu m, g = 4.5\,\mu m$	
	$d = 2\,\mu m$	$d = 4\,\mu m$	$d = 2\,\mu m$	$d = 4\,\mu m$	$d = 2\,\mu m$	$d = 4\,\mu m$
T_{max} (%)	95.6	98.8	86.2	95.2	77.5	90.3
V_{op} (V_{rms})	7.0	8.0	4.5	5.0	4.0	4.5
t_{rise} (ms)	6.8	30.2	7.0	28.0	7.6	25.7
t_{decay} (ms)	8.6	34.3	8.5	33.2	8.3	32.7

the response times of these cells roughly follow the d^2 rule, but are still about 8 ms (decay time) even for a 2-µm cell gap. By rough estimation, to obtain a response time (decay only) below 5 ms, the cell gap needs to be less than 1.6 µm. Using the same LC parameters but decreasing the cell gap to 1.6 µm with $\Delta n = 0.2$, the rise time is about 4.7 ms and the decay time is about 5.3 ms, where the peak transmittance decreases to about 74% with $w = 2$ µm and $g = 3$ µm. By simply increasing Δn to 0.25, the peak transmittance could reach about 80%. In addition, if a small amount of reverse-handed (or negative) chiral dopant is added to the LC cell to increase the maximum twist angles, another small gain in transmittance can be obtained. Such a response time is acceptable for small mobile video applications, but it is still inadequate for color sequential displays.

4.3.2 Transflective IPS and FFS Technology

Besides transmissive IPS and FFS LCDs for mobile applications, transflective-type IPS and FFS LCDs are also emerging. Unlike transflective MVA LCDs that utilize circular polarizers to achieve a common dark state for both transmissive and reflective sub-pixels, it is rather challenging to obtain a normally black reflective mode in transflective IPS or FFS cells. In the transmissive IPS or FFS sub-pixel, the LC rubbing direction is always parallel to or perpendicular to the transmission axis of the front linear polarizer to create a high contrast ratio and wide viewing angle. Thus, the reflective region is always normally white without additional treatment. As a result, for transflective IPS and FFS LCDs, the major challenge is to design a normally black reflective mode without degrading the excellent optical properties of the transmissive sub-pixel.

4.3.2.1 Transflective IPS and FFS Technology using an In-cell Retarder

To obtain a normally black reflective mode, the front linear polarizer, the retardation films (if there are any), and the LC cell in the reflective region should function together as a circular polarizer. Placing a patterned in-cell retarder (ICR) in the reflective region is a direct solution to this problem, where either half-wave or quarter-wave ICRs could be used [44–52]. Figure 4.15 shows the device configuration of a transflective FFS LCD using a half-wave ICR [44, 45]. In the transmissive region, the LC cell thickness is designed for maximum transmittance with its rubbing direction perpendicular to (or parallel to) the front linear polarizer transmission axis. In the reflective region, the LC cell is a quarter-wave plate and the ICR is a half-wave plate, which, together with the front linear polarizer, forms a broadband circular polarizer.

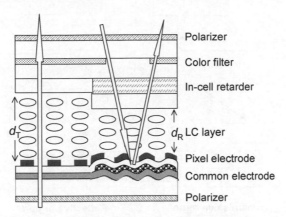

Figure 4.15 Schematic diagram of a transflective FFS cell using a half-wave ICR

On the rear substrate, patterned pixel electrode strips are formed in front of a planar common electrode to generate fringe fields; a metal layer functioning as the reflector is coated in front of the common electrode in the reflective region. To maximize the aperture, the TFT and storage capacitor can be hidden beneath the reflector. In order to boost the reflectance, the front color filters in the reflective region could have some openings as well. A tradeoff for this brightness enhancement, however, is color desaturation in the reflective mode.

The optical alignment of the device is illustrated in Figure 4.16. Clearly, the transmissive sub-pixel is like a conventional IPS or FFS cell with its $d\Delta n$ value slightly greater than 300 nm (for $\lambda = 550$ nm). In the reflective sub-pixel, the optical axis of the half-wave ICR and the rubbing direction of the quarter-wave plate LC cell are set at $-22.5°$ and $90°$ away from the front linear polarizer transmission axis, respectively. Thus, they satisfy $2\theta_{\lambda/4} - 4\theta_{\lambda/2} = \pm 90° + 2m\pi$ (here the $\pm$ sign depends on the $\theta_{\lambda/2}$ value) to enable broadband operation. As a result, a common dark state can be achieved for both transmissive and reflective modes when no voltage is applied to the LC cell. In the bright state, the transmissive sub-pixel requires the LC layer to be equivalent to a half-wave plate, and the effective averaged LC optical axis to be about $45°$ from the initial rubbing direction. Therefore, the incident backlight would experience a $90°$ rotation to pass the front linear polarizer. The reflective sub-pixel only requires the effective averaged LC optical axis to be $45°$ from its initial rubbing direction. Hence, the linearly polarized light incident from the front half-wave ICR (at $-45°$ away from the front linear polarizer transmission axis) maintains

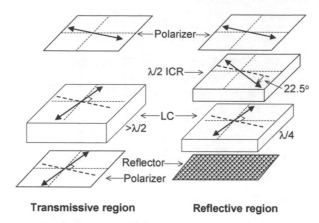

Figure 4.16 Optical configuration of a transflective FFS LCD using a half-wave ICR

linear polarization after it traverses the LC cell on to the reflector surface, and will be reflected back to transmit through the front linear polarizer.

The calculated VT and VR curves are shown in Figure 4.17. The LC material used is a positive LC material with $\Delta\varepsilon = +8.1$ and birefringence $\Delta n = 0.08$ at $\lambda = 550$ nm. The LC cell gaps for the transmissive region and reflective region are 4.0 μm and 1.72 μm, respectively. The electrode strip width and gap are 3 μm and 5 μm for both transmissive and reflective regions. For FFS cells,

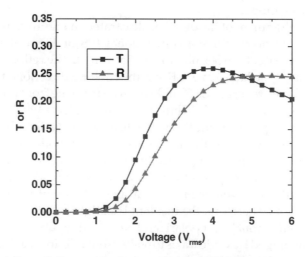

Figure 4.17 VT and VR curves of a transflective FFS LCD using a half-wave ICR and $+\Delta\varepsilon$ LC material

using a positive LC material could elicit a fast response time and low driving voltage, but the efficiency would be reduced. As a reference, the maximum transmittance from two parallel linear polarizers is about 34%. Thus, at $V = 3.75\,V_{rms}$, the normalized light efficiency for the transmissive sub-pixel and the reflective sub-pixel is relatively low at 76% and 65%, respectively. In addition, the overlap between the VT and VR curves is not good for this device configuration. For the transmissive mode, the tilt of LC directors in front of the centers of the electrode strips and gaps causes low light efficiency when a $+\Delta\varepsilon$ LC material is used. For the reflective sub-pixel, the nonuniform in-plane twist along the horizontal direction causes low light efficiency there. The discrepancy between the VT and VR curves mainly originates from the different LC cell gaps of these two regions. In the transmissive sub-pixel, the thick cell gap means the surface anchoring forces have less impact on driving voltage, leading to a maximum transmittance at $V = 3.75\,V_{rms}$. However, for the thin reflective LC cell, the more influential anchoring forces make the average LC director optical axis at $3.75\,V_{rms}$ still fall short of the optimal 45°.

To enhance reflectance, Hitachi proposed to use the ECB mode in the reflective sub-pixel rather than an FFS configuration, as shown in Figure 4.18 [46, 47]. The transmissive sub-pixel is still a wide-view FFS cell using patterned pixel electrodes and a planar common electrode on the same rear substrate to drive the LCs by fringe fields, while the reflective sub-pixel uses planar electrodes on both front and rear substrates to drive the LCs by vertical fields. The optical alignment for this configuration is the same as in the above example, but the reflective sub-pixel reaches a maximum reflectance when the LCs there

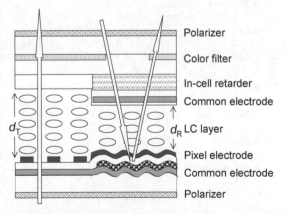

Figure 4.18 Schematic diagram of a transflective LCD using a transmissive FFS cell and a reflective ECB cell with a half-wave ICR

all tilt up to exhibit a negligible phase retardation. The previous example using reflective FFS electrodes requires the reflective LC cell to have a 45° rotation from the initial rubbing direction in the bright state. Due to surface anchoring, there is actually a dual TN profile in the bright state, which inherently limits the reflectance. In contrast, for the ECB configuration, vertical fields reorient the LCs uniformly in the whole bulk region at quite a low voltage with a large $\Delta\varepsilon$ value. As shown in Figure 4.19, even at 3.75 V_{rms}, most LCs in the reflective region are already tilted up to output a much higher reflectance ($\sim$ 31%, as compared to 22% using FFS electrodes).

In the designs above, the reflective sub-pixel utilizes a half-wave ICR in front of the LC cell. Thus, to form a circular polarizer, the LC cell thickness there needs to be adjusted to a quarter-wave plate value. Besides different cell gaps between the transmissive and reflective regions, the thin cell gap in the reflective region (if $\Delta n = 0.08$, the cell thickness is $\sim$ 1.72 μm) requires accurate thickness control. The biggest concern with the thin-cell approach is low manufacturing yield. Hence, a uniform cell gap is highly desired. To use a single-cell-gap configuration in transflective FFS LCDs, a quarter-wave plate ICR can be formed behind the LC cell in the reflective region [48–51]. The device configuration is shown in Figure 4.20, where the left-hand plot shows the electrode layout and the right-hand plot depicts the device cross-sectional view. Figure 4.21 plots the optical alignment of the device. The transmission axis of the front polarizer is aligned at −57.5° and the rear polarizer is crossed

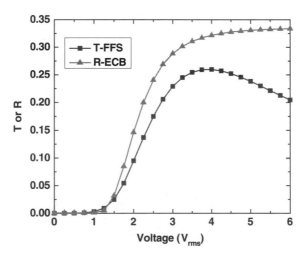

Figure 4.19 Simulated VT and VR curves of the transflective LCD using a transmissive FFS cell and a reflective ECB cell with a half-wave ICR

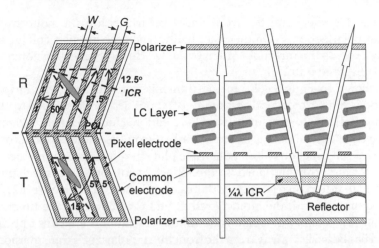

Figure 4.20 Device configuration of an FFS LCD using a quarter-wave ICR with top view of the electrode (left plot) and cross-sectional view (right plot)

to the front one. The LC rubbing directions in both regions are parallel to the front linear polarizer's transmission axis. The quarter-wave ICR has its optical axis at $-12.5°$, which is $45°$ away from the LC rubbing direction and the front linear polarizer's transmission axis. Therefore, the reflective LC cell does not change the incident light polarization in the dark state, making the

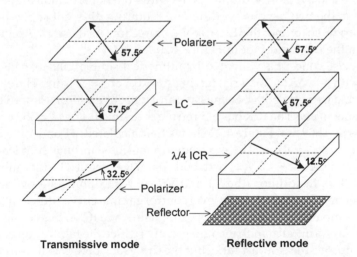

Figure 4.21 Optical configuration of the transflective FFS LCD using a quarter-wave ICR

rear ICR, together with the front linear polarizer, form a monochromatic circular polarizer only. On the other hand, when the applied voltage significantly exceeds the threshold voltage, the LC directors are reoriented by the electric fields to output light.

As shown in Figure 4.20, in the transmissive sub-pixel, the strip pixel electrodes are formed with an angle of about 15° with respect to the LC rubbing direction to gain an optimal LC rotation. Under proper reorientation (its effective optical axis is about 45° away from its initial orientation), the whole LC cell functions like a half-wave plate to rotate the rear incident light by 90° for high transmittance. In the reflective sub-pixel, the mechanism to obtain high reflectance is slightly more complicated. From Figure 4.21, it has two requirements: (i) the on-state reflective LC cell needs to be functionally equivalent to a half-wave plate, and (ii) the on-state LC effective optical axis should be about 22.5° away from the front linear polarizer's transmission axis. Thus, the incident linearly polarized light from the front polarizer will be rotated 45° by the half-wave-like reflective LC cell to become perpendicular to the optical axis of the ICR. Hence, the ICR will not change the polarization of the incident and reflected light passing through it, and the reflected light experiences a second 45° rotation by the LC cell in front and becomes parallel to the front polarizer transmission axis.

From the above analysis, both regions require the LC cell to function as a half-wave plate for optimal light output in the bright state, thus a single cell gap can be achieved. In addition, the required on-state rotation angles of the LC layer in the two regions are different, requiring different angles between the electrode strip and the LC rubbing direction in these two regions, as shown in the left-hand plot in Figure 4.20.

The device concept is validated by numerical simulations using a 3.5-μm LC cell with a $+ \Delta \varepsilon$ LC material having $\Delta n = 0.1$ at $\lambda = 550$ nm. The electrode width w and gap g for both transmissive and reflective sub-pixels are 3 μm and 4 μm, respectively. The ICR has a birefringence $\Delta n = 0.17$ at $\lambda = 550$ nm and a thickness of 0.809 μm. Figure 4.22 shows the calculated VT and VR curves of the device at $\lambda = 550$ nm. The electrode dimensions and material lead to an operating voltage at $\sim 4.5\,V_{rms}$ with both T and R at about 28% (of a maximum value of 34%). In addition, the VT and VR curves exhibit good overlap with each other, making a single gray level control gamma curve sufficient for both modes. Compared with the designs using half-wave ICRs, this configuration utilizes a uniform cell gap that could greatly reduce the fabrication complexity and improve the contrast ratio. But the drawback here is that the reflective mode only obtains a good dark state at one central wavelength (say, 550 nm). For a transflective LCD, the performance of the transmissive sub-pixel is

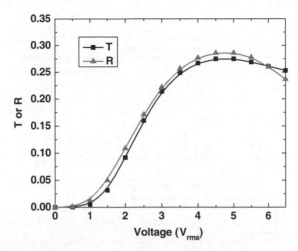

Figure 4.22 Calculated VT and VR curves of the transflective FFS LCD using a quarter-wave ICR behind the LC cell

important and the reflective sub-pixel is less significant. Of course, better solutions for obtaining high transmittance and high reflectance simultaneously in a single device are desirable.

The transflective FFS LCD examples discussed above all require a patterned in-cell retarder in the reflective sub-pixel to obtain a common dark state for both transmissive and reflective modes. The major technical challenge lies in the difficult fabrication process of the patterned retarder in mobile displays, where pixel dimension is quite small ($< 50\,\mu m$ including both T and R sub-pixels). Figure 4.23 illustrates the fabrication process of patterned ICRs [52]. First, an alignment layer is formed on the substrate surface and treated with rubbing or other alignment means for the required orientation, then a UV-curable LC material (such as LC monomers with a large Δn value for thin cell thickness) is coated on the alignment layer. Next, the substrate is irradiated with UV light through patterned masks to polymerize the LCs. Typically, each polymerized LC region has a trapezoidal shape in the cross-section from the UV irradiation through masks. Finally, the LC in the unexposed area is developed away to generate the designed patterns. To form a high-quality ICR with an accurate retardation value, small dimension, square edges, and good thermal stability, several factors such as UV dosage, distance between the mask and the LC surface, and development condition all need to be considered [52]. Presently, high-quality patterned ICRs smaller than $15\,\mu m$ with a thickness less than $2\,\mu m$ are obtainable for high-resolution

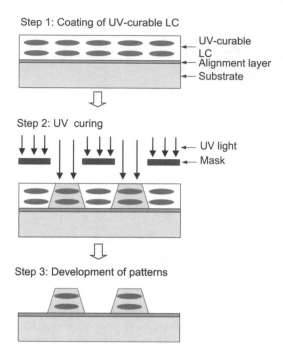

Step 1: Coating of UV-curable LC

UV-curable
LC
Alignment layer
Substrate

Step 2: UV curing

UV light
Mask

Step 3: Development of patterns

Figure 4.23 Fabrication process of an in-cell retarder (redrawn from (52))

transflective LCD applications. Nevertheless, the fabrication of patterned ICRs is still not easy to control, considering the complex treatments such as buffing the alignment layers, UV curing through fine mask slits, and development process, not to mention the delicate TFT and color filter formation processes. These factors result in practical challenges for fabricating ICR-embedded transflective FFS LCDs with a high yield. Therefore, in parallel to perfecting ICR fabrication technologies, researchers are also actively developing ICR-free transflective FFS LCDs.

4.3.2.2 Transflective IPS and FFS Technology without an In-cell-retarder

In this section, several designs for ICR-free transflective FFS LCDs are reviewed. The design consideration is similar to the above examples; that is, primarily to attain inherent high performance of the transmissive mode and to obtain an acceptable normally black reflective mode. Each design has its own merits and demerits.

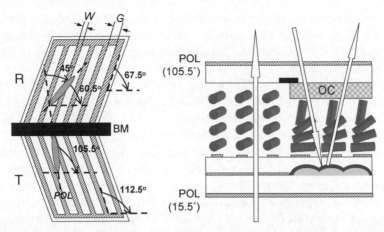

Figure 4.24 Transflective LCD using a transmissive FFS cell and a reflective HAN cell

A first method to remove the patterned ICR from transflective FFS LCDs is to employ dual surface alignment. Figure 4.24 depicts the device configuration of a transflective LCD combining a transmissive FFS cell and a reflective hybrid aligned nematic (HAN) cell [53]. In the transmissive region, the LC rubbing directions on both substrates are parallel to the front polarizer's transmission axis to ensure good performance in the transmissive sub-pixel, but the alignment directions of the reflective HAN cell on the substrates are set at 45° with respect to the front linear polarizer's transmission axis. For the dark state in the reflective sub-pixel, the 45° aligned (with respect to the polarizer's transmission axis) HAN cell functions like a quarter-wave plate at normal incidence. The different surface alignments could be obtained by applying rubbing in the transmissive region and photo-alignment in the reflective region, or by applying photo-alignment in both regions. The main reason to select a HAN cell is to make the cell gaps in the transmissive and reflective regions similar, since the effective phase retardation of a HAN cell is only about one half of its assigned $d\Delta n$ value. From simulation, it is found that by selecting a $+\Delta\varepsilon$ LC material, to get sufficient transmittance, the $d\Delta n$ value in the transmissive region needs to be significantly higher than 0.30 μm (such as 0.38 μm), thus an overcoat (OC) layer ($\sim$ 1 μm thick) is needed in front of the reflective LC cell in order to adjust the cell thickness there to obtain a quarter-wave retardation. However, when a $-\Delta\varepsilon$ LC material is used, the in-plane rotation of LC directors is much more uniform than that using a $+\Delta\varepsilon$ material. Thus, a smaller cell gap in the transmissive region can be adopted to obtain sufficient transmittance, which, in turn, could reduce the thickness of the OC layer or even eliminate it. The tradeoff is the increased response time because a $-\Delta\varepsilon$ LC material usually possesses a higher viscosity than its positive counterpart.

The device concept is validated by simulation, where the LC material selected has a $\Delta\varepsilon = +9.3$ and $\Delta n = 0.105$ at $\lambda = 550$ nm. The LC cell gaps for the transmissive region and reflective region are 3.62 μm and 2.74 μm, respectively. In other words, the OC layer is about 0.88 μm. To obtain a bright state requires the transmissive LC cell to be equivalent to a 45° rotated half-wave plate, and the reflective LC layer 45° rotated from its initial rubbing direction. The angles between the initial LC director alignment and the edge of the electrode strips are both 7°. The calculated VT and VR curves with the above settings are shown in Figure 4.25, where the maximum transmittance and reflectance at $V = 4.5\,V_{rms}$ are about 28% and 29% (of a maximum possible value of 34%), respectively. To improve the reflectance, a HAN cell can be switched between a planar common electrode on the front substrate, and a planar pixel electrode on the rear, which uses vertical fields to perturb the LC directors. Unlike the method using ICRs, this device uses dual alignment in the transmissive and reflective regions to achieve a normally black reflective mode. Presently, the control of pixelized dual alignment is still quite challenging. But as the photo-alignment technique improves, the dual-alignment

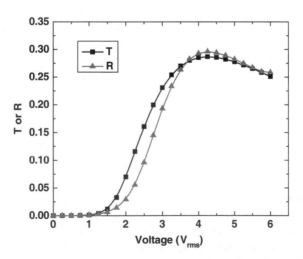

Figure 4.25 Simulated VT and VR curves of a transflective FFS LCD using dual alignment

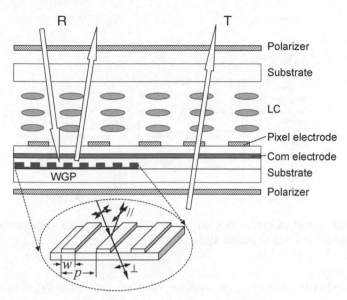

R   T

Polarizer
Substrate
LC
Pixel electrode
Com electrode
Substrate
Polarizer

WGP

w
p

Figure 4.26 Transflective LCD using a transmissive FFS cell with a WGP as reflector

approach may become feasible for realizing high-performance transflective FFS LCDs for mobile applications.

Another approach is to use a nano-wire grid polarizer (WGP) in each pixel as the polarization-dependent reflector [54]. The device configuration is shown in Figure 4.26, where transmissive and reflective regions share the same LC cell thickness. As shown in the dashed circle, the WGP functions as a polarization-dependent reflector. For unpolarized incident light, it reflects the light with its polarization parallel to the metal ribs, and transmits the other component polarized perpendicular to the grids [55–57]. Recently, WGPs for visible light have become available by using UV-nanoimprint technology [58–60], and pixelized WGPs on TFT glass substrates have already been applied for stereoscopic displays [61]. These developments make the manufacture of WGPs in transflective LCD devices achievable.

The working mechanism of the transflective LCD is illustrated in Figure 4.27. The LC rubbing direction is parallel to (or perpendicular to) the front linear polarizer's transmission axis and the WGP has its metal grids perpendicular to the front linear polarizer's transmission axis. The switching principle for the transmissive sub-pixel is the same as a transmissive FFS LCD. For the reflective sub-pixel, in the dark state, the LCs remain unperturbed, thus the incident

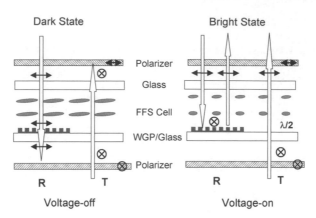

Figure 4.27 Operating mechanisms of a WGP-embedded transflective LCD in the dark state (left) and bright state (right)

ambient light maintains its polarization when it reaches the WGP surface, it is then transmitted by the WGP and is further absorbed by the rear linear polarizer. In consequence, a normally black reflective mode is obtained. On the other hand, once a relatively high voltage is applied, the LCs in both transmissive and reflective regions are tuned to behave like a 90° polarization rotator (such as a half-wave plate aligned at 45° from the incident light polarization direction), thus the incident ambient light with its polarization parallel to the wire grids is reflected back to the LC cell. Upon reflection, the light experiences a second 90° polarization rotation by the LC layer, thus transmitting through the front linear polarizer.

However, at an intermediate gray level where the effective phase retardation of the LC layer is less than the half-wave value, the optical output from the transmissive and reflective regions is quite different. For the transmissive sub-pixel, the backlight traverses the LC cell once and becomes elliptically polarized, so that only the component parallel to the front linear polarizer's transmission axis can pass to the viewer. For the reflective sub-pixel, additional loss of light will occur at the WGP surface. Incident ambient light reaching the WGP surface has an elliptical polarization that has components both parallel and perpendicular to the wire grids. Only the component polarized parallel to the metal grids will be reflected back to the LC cell, while the remaining portion is transmitted by the WGP and gets lost. If we compare the reflected light from the WGP surface with the backlight in the transmissive sub-pixel, we find that their optical paths are identical to each other. If we compare the optical paths of the reflected light from the

WGP surface to the front polarizer with light propagating from the backlight in the transmissive sub-pixel from the rear polarizer front surface to the front polarizer, we find that they are identical to each other and will experience a similar intensity loss. In other words, the previous loss during the incidence on the WGP surface is additional for the reflected light. Consequently, the reflective mode exhibits a higher threshold voltage than the transmissive mode, and the normalized VT and VR curves will diverge from each other. To solve this problem, we can tune the threshold voltage of the transmissive sub-pixel a little higher than the reflective sub-pixel, which can be achieved by using different electrode width w and gap g in the transmissive and reflective regions.

To simulate device performance, we use FEM to compute the LC director profile and the 4×4 matrix method [54] to calculate the LC optics (AR coating is used to remove the surface reflection). Both the electrode width w and electrode gap g in the transmissive region are set at $3\,\mu m$, and $w = 3\,\mu m$ and $g = 5\,\mu m$ are chosen for the reflective region. A finer electrode pattern usually produces a higher threshold voltage and on-state voltage in an FFS cell. Figure 4.28 shows the calculated VT and VR curves. Here, the small amount of light leakage in the VR curve below the threshold voltage comes from the imperfect AR coating employed in the 4×4 matrix method calculation. To reduce the driving voltage, the passivation layer between pixel and common electrodes could also be optimized.

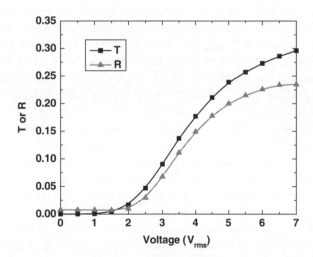

Figure 4.28 Simulated VT and VR curves for the WGP-embedded transflective FFS LCD

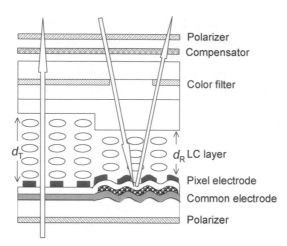

Figure 4.29 Device configuration of a film-compensated transflective FFS LCD

Besides using patterned ICRs, dual alignment, or patterned nano WGP, another method is to employ compensation films such as a negative uniaxial A-film to obtain a common dark state between the transmissive and reflective sub-pixels [62, 63]. Figure 4.29 shows the proposed device configuration where a uniaxial negative A-film is placed in front of an FFS LC cell to fully compensate the phase retardation of the transmissive LC cell (which is like a positive A-film in the dark state) and at the same time to function together with the reflective LC cell as a quarter-wave plate. In the figure, the initial LC rubbing directions in both regions are set at 45° from the transmission axis of the front linear polarizer. First, the transmissive LC cell gap d_T is selected to have an optimal transmittance in the voltage-on state (say: $d_T\Delta n > 300$ nm). Then, the in-plane retardation of the compensator is designed to fully compensate the transmissive LC sub-pixel. The cell gap d_R in the reflective region is then selected together with the compensator to have an overall retardation value as a quarter-wave plate at the designed wavelength, say 550 nm. Thus, the front linear polarizer, the compensator, and the reflective LC cell form a circular polarizer to obtain a dark state when no voltage is applied. For the transmissive sub-pixel, a compensator like a negative A-film would fully cancel the phase retardation at both normal and off-axis incidence; thus, a wide viewing angle transmissive sub-pixel is also obtained. Using real LC material and compensation film data (negative A-film) in device simulation, we obtain good light efficiency for both transmissive and reflective modes. The calculated VT and VR curves are

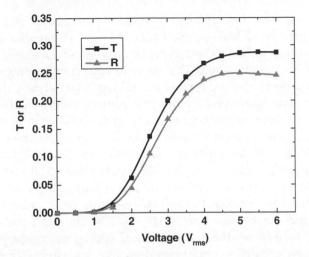

Figure 4.30 Simulated VT and VR curves for the transflective FFS LCD using a uniform compensator

shown in Figure 4.30, where a good match is obtained between the normalized VT and VR curves. In this device, the compensator is a uniform plate covering both transmissive and reflective regions, but the LC cell gaps are still different in the transmissive and reflective regions. However, the cell gap of the reflective region could be adjusted by the phase retardation value of the transmissive sub-pixel, thus it could reach over 2 μm for ease of fabrication. In comparison, in the other dual-cell-gap designs, the reflective cell gap value needs to be about a quarter-wave plate, thus making the thickness quite small ($\sim 1.5\,\mu m$).

In addition to the above designs using in-cell retarders or other means like wire grid polarizers, there are other interesting approaches for realizing transflective IPS or FFS LCDs, such as using an IPS electrode to drive TN cells [64], using the biased reflective mode [65]. Each method has its unique merits and demerits.

4.4 Summary

In this chapter, some wide-view technologies, including multi-domain vertical alignment (MVA), in-plane switching (IPS), and fringe-field switching (FFS) LCDs, have been briefly reviewed. MVA technologies utilize either protrusions or slits, or both, as the means of guiding LC directors to form multi-domains; thus no rubbing treatment is necessary and the device exhibits an inherently high contrast ratio. MVA technologies for different panel applications have different requirements and features. For large-panel applications like monitors and TVs, four-domain configurations using zig-zag-patterned electrodes are typically employed. To reduce color washout, eight-domain configurations are not uncommon, where the transmittance dependence on both azimuthal angle and polar angle can be reduced. To improve the dynamic response of MVA LCDs, jagged electrodes have been developed to extend the guiding effect along the horizontal direction. In addition, polymer-sustained surface alignment (PSA) technology has been developed to introduce a small pre-tilt angle within each domain, which could significantly enhance the device response time and transmittance. For small-panel devices like mobile MVA LCDs, high transmittance is very important. To improve the transmittance, special electrode patterns are designed to have axially continuous domains and circular polarizers are therefore used. But the color washout can be severe in mobile MVA cells using circular polarizers. Here, PSA technology is greatly beneficial to MVA LCDs, with a better dynamic performance and resistance to external pressure for touch panels. For IPS- or FFS-based LCDs, the pixel design considerations have been briefly discussed, including methods to achieve high aperture ratio, low crosstalk, and high transmittance for a given aperture. For IPS or FFS LCDs, the color gamma shift is much less than for MVA LCDs, and the viewing angle is inherently wide even without any compensation film, making them very attractive for mobile displays. Much effort is being made to further improve the transmittance and suppress the color shift. Cell designs to achieve transflective IPS or FFS LCDs have also been reviewed in detail, including the device configurations and electro-optical properties of the designs using in-cell retarders (ICRs) and ICR-free configurations using dual alignment, wire grid polarizers, or compensation films. The merits and demerits of each technology have also been addressed. As fabrication technology continues to advance, the widespread adoption of MVA, IPS, and FFS technologies into mobile displays is foreseeable.

References

[1] Takeda, A., Kataoka, S., Sasaki, T., Chida, H., Tsuda, H., Ohmuro, K., Sasabayashi, T., Koike, Y. and Okamoto, K. (1998) A super-high image quality multi-domain vertical alignment LCD by new rubbing-less technology. *SID Tech. Digest*, **29**, 1077–1080.

[2] Tanaka, Y., Taniguchi, Y., Sasaki, T., Takeda, A., Koibe, Y. and Okamoto, K. (1999) A New Design to Improve Performance and Simplify the Manufacturing Process of High-Quality MVA TFT-LCD Panels. *SID Tech. Digest*, **30**, 206–209.

[3] Lien, S.C. and John, R.A. (1994) Liquid Crystal Displays Having Multi-Domain Cells. U.S. Patent 5 309 264 (May 1994).

[4] Lien, A. and John, R.A. (1993) Multi-Domain Homeotropic Liquid Crystal Display for Active Matrix Application. *EuroDisplay '93*, p. 21.

[5] Lien, S.C.A., Cai, C., Nunes, R.W., John, R.A., Galligan, E.A., Colgan, E. and Wilson, W.S. (1998) Multi-domain homeotropic liquid crystal display based on ridge and fringe field structure. *Jpn J. Appl. Phys.*, **37**, L597.

[6] Kim, K.H., Lee, K.H., Park, S.B., Song, J.K., Kim, S.N. and Souk, J.H. (1998) Domain divided vertical alignment mode with optimized fringe field effect. In *Proc. 18th Int. Display Research Conf.* (Asia Display'98), pp. 383–386.

[7] Kataoka, S., Takeda, A., Tsuda, H., Koike, Y., Inoue, H., Fujikawa, T., Sasabayashi, T. and Okamoto, K. (2001) A New MVA-LCD with Jagged Shaped Pixel Electrodes. *SID Tech. Digest*, **32**, 1066–1069.

[8] Wei, C.K. (2008) The progress of TFT-LCD. Keynote speech, IDRC'08, Orlando.

[9] Song, J.-K., Lee, K.-E., Chang, H.-S., Hong, S.-M., Jun, M.-B., Park, B.-Y., Seomun, S.-S., Kim, K.-H. and Kim, S.-S. (2004) DCCII: Novel method for fast response time in PVA mode. *SID Tech. Digest*, **35**, 1344–1347.

[10] Hanaoka, K., Nakanishi, Y., Inoue, Y., Tanuma, S. and Koike, Y. (2004) A New MVA-LCD by Polymer Sustained Alignment Technology. *SID Tech. Digest*, **35**, 1200–1203.

[11] Kim, S.G., Kim, S.M., Kim, Y.S., Lee, H.K., Lee, S.H., Lee, G.-D., Lyu, J.-J. and Kim, K.H. (2007) Stabilization of the liquid crystal director in the patterned vertical alignment mode through formation of pretilt angle by reactive mesogen. *Appl. Phys. Lett.*, **90**, 261910–261912.

[12] Hsu, S.-F., Chien, W.-Y., Wu, K.-H., Yeh, L.-Y., Wu, M.-H., Lin, T.C. and Tsai, J.-C. (2008) Advanced-MVA mobile technology for fast-switching LCD displays. *SID Tech. Digest*, **39**, 503–506.

[13] Chen, T.-J. and Chu, K.-L. (2008) Pretilt angle control for single-cell-gap transflective liquid crystal cells. *Appl. Phys. Lett.*, **92**, 091102.

[14] Lee, S.H., Kim, S.M. and Wu, S.T. (2009) Emerging vertical alignment liquid crystal technology associated with surface modification using UV curable monomer. *J. Soc. Info. Disp.*, **17**, 551–559.

[15] Lee, Y.-J., Kim, Y.-K., Jo, S.I., Gwag, J.S., Yu, C.-J. and Kim, J.-H. (2009) Surface-controlled patterned vertical alignment with reactive mesogen. *Opt. Express*, **17**, 10298.

[16] Chen, T.-S., Huang, C.-W., Hsieh, C.-C., Huang, B.-H., Jian, J.-H., Chan, T.-W., Tsao, C.-H., Chen, C.-W., Chen, C.-Y., Chiu, C.-H., Pai, C.-H., Jaw, J.-H., Lin, H.-C., Chiu, C.-Y., Su, J.-J., Norio, S., Chang, T.-J., Liau, W.-L. and Lien, A. (2009) Advanced-MVA III technology for high-quality LCD TVs. *SID Tech. Digest*, **40**, 776–779.

[17] Shih, P.-S., Chen, K.-T., Wang, W.-H., Pan, H.-L., Chen, P.-Y., Lin, C.-Y., Lin, S.-H. and Yang, K.-H. (2006) A new pixel design and a novel driving scheme for multi-domain vertically-aligned LCDs. *SID Tech. Digest*, **37**, 1067–1070.

[18] Kim, S.S., Berkeley, B.H., Park, J.H. and Kim, T. (2006) New era for TFT-LCD size and viewing-angle performance. *J. Soc. Info. Disp.*, **14**, 127.

[19] Tai, M.-C., Chang, M.-H., Liu, C.-C., Chang, Y.-P. and Wang, M.-T. (2007) MVA-LCD with low color shift and high image quality. *SID Tech. Digest*, **38**, 1007–1009.

[20] Huang, Y.-P. (2007) Additional refresh technology (ART) of advanced-MVA (AMVA) mode for high quality LCDs. *SID Tech. Digest*, **38**, 1010–1013.

[21] Shih, P.-S., Lin, J.-S., Pan, H.-L., Chen, P.-Y., Liao, T.-S. and Yang, K.-H. (2008) Development of asymmetric-gate-coupled eight domain (AGC-8D) HVA for TFT-LCD TVs. *SID Tech. Digest*, **39**, 208–211.

[22] Kim, S.S., You, B.H., Cho, J.H., Moon, S.J., Berkeley, B.H. and Kim, N.D. (2008) 82″ Ultra definition LCD using new driving scheme and advanced super PVA technology. *SID Tech. Digest*, **39**, 196–199.

[23] Lu, Y.-C., Cheng, T.-C., Lee, S.-L., Hu, C.-J. and Gan, F.-Y. (2007) Color washout improvement for mobile display application. *IDW'07*, pp. 1707–1710.

[24] Bae, K.-S., Gwag, J.S., Han, I.-Y., Lee, Y.-J., Choi, Y., Choi, J.-S. and Kim, J.-H. (2008) Azimuthally continuous nematic domain mode using electrode structure with circular slit. *SID Tech. Digest*, **39**, 1951–1954.

[25] Kim, J.H., Park, W.S., Yeo, Y.S., Lee, J.Y., Ahn, S.H. and Kim, C.W. (2007) Novel PVA pixel design for mobile touch screen panel application with excellent optical performance and pressure resistant characteristics. *IDW'07*, pp. 1723–1726.

[26] Shibazaki, M., Sugiyama, M., Takahashi, S., Yoshiga, M., Inada, T., Lee, H.-C., Yu, M.-W., Wu, M.-F., Chang, C.-H., Chang, T.-S., Chang, Y.-J., Wu, I.-L., Chang, W.-C. and Ting, D.-L. (2007) MVA mode with improved color-wash-out for mobile application. *SID Tech. Digest*, **38**, 1665–1668.

[27] Lin, C.-H., Huang, K.-Y. and Lin, H.-Y. (2006) High transmittance MVA-LCD with low color shift. *IDW'06*, pp. 137–140.

[28] Kang, S.-G., Kim, S.-H., Song, S.-C., Park, W.-S., Yi, C., Kim, C.-W. and Chung, K.H. (2004) Development of a Novel Transflective Color LTPS-LCD with Cap-Divided VA-Mode. *SID Tech. Digest*, **35**, 31–33.

[29] Yang, Y.-C., Choi, J.Y., Kim, J., Han, M., Chang, J., Bae, J., Park, D.-J., Kim, S.I., Roh, N.-S., Kim, Y.-J., Hong, M. and Chung, K. (2006) Single Cell Gap Transflective Mode for Vertically Aligned Negative Nematic Liquid Crystals. *SID Tech. Digest*, **37**, 829–831.

[30] Ge, Z., Zhu, X., Lu, R., Wu, T.X. and Wu, S.T. (2007) Transflective liquid crystal display using commonly biased reflectors. *Appl. Phys. Lett.*, **90**, 221111.

[31] Lu, R., Ge, Z. and Wu, S.T. (2008) Wide-view and single cell gap transflective liquid crystal display using slit-induced multidomain structures. *Appl. Phys. Lett.*, **92**, 191102.

[32] Lin, C.H., Chen, Y.R., Hsu, S.C., Chen, C.Y., Chang, C.M. and Lien, A. (2008) A novel advanced wide-view transflective display. *J. Disp. Technol.*, **4**, 123.

[33] Bauer, G., Kiefer, R., Klausmann, H. and Windscheid, F. (1995) In-plane switching: A novel electro-optic effect. *Liq. Cryst. Today*, **5**, 12–13.

[34] Soref, R.A. (1974) Field effects in nematic liquid crystals obtained with interdigital electrodes. *J. Appl. Phys.*, **45**, 5466.

[35] Ohe, M. and Kondo, K. (1995) Electro-optical characteristics and switching behavior of the in-plane switching mode. *Appl. Phys. Lett.*, **67**, 3895.

[36] Lee, S.H., Lee, S.L. and Kim, H.Y. (1998) Electro-optic characteristics and switching principle of a nematic liquid crystal cell controlled by fringe-field switching. *Appl. Phys. Lett.*, **73**, 2881.

[37] Yamakita, H., Nishiyama, K., Shiota, A., Ogawa, S., Kumagawa, K. and Takimoto, A. (2001) Electro-Optical Characteristics of In-Plane Switching LCDs using Transparent Electrodes Structure. *IDW'01*, pp. 213–216.

[38] Nakayoshi, Y., Kurahashi, N., Tanno, J., Nishimura, E., Ogawa, K. and Suzuki, M. (2003) High transmittance pixel design of in-plane switching TFT-LCDs for TVs. *SID Tech. Digest*, **34**, 1100–1103.

[39] Ge, Z., Zhu, X., Wu, T.X. and Wu, S.T. (2006) High transmittance in-plane switching liquid crystal displays. *IEEE/OSA J. Disp. Technol.*, **2**, 114–120.

[40] Lee, S.H., Lee, S.L., Kim, H.Y. and Eom, T.Y. (1999) A novel wide-viewing-angle technology: ultra-trans view™. *SID Tech. Digest*, **30**, 202–205.

[41] Lin, J.-S., Yang, K.-H. and Chen, S.-H. (2004) A high-aperture-ratio and low-crosstalk pixel structure design for in-plane-switching-mode TFT-LCDs. *J. Soc. Inf. Disp.*, **12**, 533.

[42] Lin, J.-S., Shih, P.-S., Liao, T.-S., Chang, M.-C., Lin, R.-H., Yang, K.-H., Lin, P.-H., Wu, C.-T. and Huang, M.-P. (2007) A novel pixel structure to eliminate thick organic overcoat on the array substrate of AS-IPS mode for TFT-LCDs. *SID Tech. Digest*, **38**, 1171–1174.

[43] Lin, Y.-C., Wang, C.-Y., Chi, C.-Y., Chen, C.-J., Yang, C.-L. and Pang, J.-P. (2006) A novel pixel with small color shift for fringe field switching mode LCD. *IDW'06*, pp. 149–152.

[44] Tanno, J., Morimoto, M., Igeta, K., Imayama, H., Komura, S. and Nagata, T. (2006) A new transflective IPS-LCD with high contrast ratio and wide viewing angle performance. *IDW'06*, pp. 635–638.

[45] Imayama, H., Tanno, J., Igeta, K., Morimoto, M., Komura, S., Nagata, T., Itou, O. and Hirota, S. (2007) Novel pixel design for a transflective IPS-LCD with an in-cell retarder. *SID Tech. Digest*, **38**, 1651–1654.

[46] Hirota, S., Oka, S., Itou, O., Igeta, K., Morimoto, M., Imayama, H., Komura, S. and Nagata, N. (2007) Transflective LCD combining transmissive IPS and reflective in-cell retarder ECB. *SID Tech. Digest*, **38**, 1661–1664.

[47] Koma, N., Mitsui, M., Tanaka, Y. and Endo, K. (2007) A transflective in-plane-switching LCD with a high-optical-performance reflective area. *SID Tech. Digest*, **38**, 1270–1273.

[48] Park, J.B., Kim, H.Y., Jeong, Y.H., Kim, S.Y. and Lim, Y.J. (2005) Novel transflective display with fringe-field switching mode. *Jpn. J. Appl. Phys.*, **44**, 7524–7527.

[49] Lee, G.S., Kim, J.C., Yoon, T.-H., Kim, Y.S., Kang, W.S. and Park, S.I. (2006) Design of wide-viewing-angle transflective IPS LCD. *26th Proc. International Display Research Conference*, pp. 75–77.

[50] Lim, Y.J., Lee, M.H., Lee, G.D., Jang, W.G. and Lee, S.H. (2007) A single-gap transflective fringe field switching display using a liquid crystal with positive dielectric anisotropy. *J. Phys. D: Appl. Phys.*, **40**, 2759.

[51] Ge, Z., Wu, S.T. and Lee, S.H. (2008) Wide-view and sunlight readable transflective liquid crystal display for mobile applications. *Opt. Lett.*, **33**, 2623.

[52] Hasebe, H., Kuwana, Y., Nakata, H., Yamazaki, O., Takeuchi, K. and Takatsu, H. (2008) High quality patterned retarder for transflective LCDs. *SID Tech. Digest*, **39**, 1904–1907.

[53] Kim, H.Y., Ge, Z., Wu, S.T. and Lee, S.H. (2007) Wide-view transflective liquid crystal display for mobile applications. *Appl. Phys. Lett.*, **91**, 231108.

[54] Ge, Z., Wu, T.X. and Wu, S.T. (2008) Single cell gap and wide-view transflective liquid crystal display using fringe field switching and embedded wire grid polarizer. *Appl. Phys. Lett.*, **92**, 051109.

[55] Yu, X.J. and Kwok, H.S. (2003) Optical wire-grid polarizers at oblique angles of incidence. *J. Appl. Phys.*, **93**, 4407.

[56] Perkins, R.T., Hansen, D.P., Gardner, E.W., Thorne, J.M. and Robbins, A.A. (2000) Broadband wire grid polarizer for the visible spectrum. U.S. Patent 6 122 103, September.

[57] Ge, Z. and Wu, S.T. (2008) Nano-wire grid polarizer for energy efficient and wide-view liquid crystal displays. *Appl. Phys. Lett.*, **93**, 121104.

[58] Wang, J., Walters, F., Liu, X., Sciortino, P. and Deng, X. (2007) High-performance, large area, deep ultraviolet to infrared polarizers based on 40 nm line/78 nm space nanowire grids. *Appl. Phys. Lett.*, **90**, 061104.

[59] Wang, J., Chen, L., Liu, X., Sciortino, P., Liu, F., Walters, F. and Deng, X. (2006) 30-nm-wide aluminum nanowire grid for ultrahigh contrast and transmittance polarizers made by UV-nanoimprint lithography. *Appl. Phys. Lett.*, **89**, 141105.

[60] Ahn, S.H., Kim, J.-S. and Guo, L.J. (2007) Bilayer metal wire-grid polarizer fabricated by roll-to-roll nanoimprint lithography on flexible plastic substrate. *J. Vac. Sci. Technol. B.*, **25**, (6), 2388.

[61] Oh, J.H., Kang, D.H., Park, W.H., Kim, H.J., Hong, S.M., Hur, K.H., Jang, J., Lee, S.J., Kim, M.J., Kim, S.K., Park, K.H., Gardner, E., Hansen, J., Yost, M. and Hansen, D. (2007) High-Resolution Stereoscopic TFT-LCD with Wire Grid Polarizer. *SID Tech. Digest*, **38**, 1164–1167.

[62] Ge, Z., Wu, S.T., Li, W.Y. and Wei, C.K. (2009) Wide viewing angle transflective liquid crystal displays. U.S. patent pending, February.

[63] Matsushima, J., Uehara, S. and Sumiyoshi, K. (2007) Novel transflective IPS-LCDs with three retardation plates. *IDW'07*, pp. 1511–1514.

[64] Itou, O., Hirota, S., Sekiguchi, Y., Komura, S., Morimoto, M., Tanno, J., Fukuda, K., Ochiai, T., Imayama, H., Nagata, T. and Miyazawa, T. (2006) A wide viewing angle transflective IPS LCD applying new optical design. *SID Tech. Digest*, **37**, 832–835.

[65] Ochiai, T., Sasaki, T., Miyazawa, T., Maki, M., Morimoto, M. and Ohkura, M. (2007) Low cost retarder-less transflective IPS-LCD. *SID Tech. Digest*, **38**, 1258–1261.

5

Color Sequential Mobile LCDs

5.1 Overview

Color sequential technology, also known as field sequential color, is becoming popular as a method for eliminating the color filters in LCDs to enhance their light efficiency [1–9]. To make color sequential direct-view LCDs, the backlight sequentially emits red (R), green (G), and blue (B) light, and the LCD panel is synchronized with the backlight to display the required gray levels of each color. Besides the tremendous gain (up to 3x) in light transmittance, advantages of color sequential technology also include a wide color gamut and high resolution. Small-panel LCDs using a white LED backlight typically produce about 50% of the NTSC color gamut, but this can be raised to over 90% by using RGB LEDs to display more saturated colors. In addition, merging three R, G, and B sub-pixels into a larger pixel greatly eases the fabrication of high-resolution displays. However, compared with conventional LCDs, color sequential displays face several challenges. First, color sequential technology requires the LC cell to have a much faster response time. For example, for a regular frame refresh rate of 60 Hz, each color field changes three times faster (at 180 Hz or within 5.56 ms) and the optical response time of the LC cell needs to be less than $\sim$2 ms. A slower response time would result in a lower transmittance.

Achieving a fast response time for nematic LCs without sacrificing other performance criteria, such as transmittance, viewing angle, and manufacturing

Transflective Liquid Crystal Displays   Zhibing Ge and Shin-Tson Wu
© 2010 John Wiley & Sons, Ltd

yield, is by no means an easy task. In color sequential LCDs, especially in large panels, color breakup can be observed. Color breakup originates from mismatched movement between the human retina and moving targets, which generates undesired edge-color-blurred moving images [10–14]. Color breakup is a serious issue for large-panel color sequential LCD TVs and requires complex solutions, such as increasing the refresh rate, using motion compensation, or other techniques [10–14]. Fortunately, for mobile displays with a relatively small panel size, color breakup appears much less significant and is more tolerable to viewers. Nevertheless, the increase in brightness by the removal of color filters and low power consumption make color sequential technology quite attractive for the next generation of displays.

In this chapter we first cover the typical driving schemes of color sequential displays, and then briefly investigate the major fast-response LC modes suitable for color sequential technology, such as thin TN cells [9], pi-cells [15], and OCB (optically compensated bend) cells [16]. Fast-response transflective LCDs using conventional pi-cells or OCB cells with color filters and color sequential technology will also be addressed.

5.2 Color Sequential Driving Schemes

To drive active matrix color sequential LCDs, the driving scheme of scanning data, the LC response time, and backlight flashing time are all quite important. Figure 5.1 illustrates a typical driving scheme for a color sequential

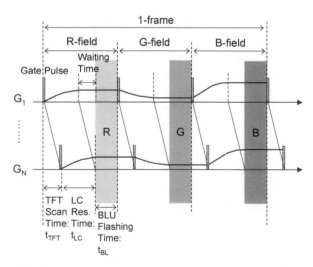

Figure 5.1 Typical driving scheme of a color sequential display

display. Each image frame at a scanning frequency of f (e.g., $f = 60\,Hz$) is equally divided into R, G, and B sub-fields, each with a time of $1/(3f)$. Within one sub-field period, the gate lines are scanned sequentially by a short gate pulse from G_1 to G_N, which occupies a total scan time of t_{TFT}. On application of the driving voltage from the TFT, each LC pixel takes time t_{LC} to finish the gray-level transitions. When the last LC pixel reaches its optical saturation level, the backlight of the assigned color (say, R) flashes for a time period of t_{BL} until the next sub-field (say, G) starts. Here, in the plot, for a simple illustration, the transmittance curve is assumed to be the same for all pixels in each sub-field; in reality, this could vary with practical image data.

From the above analysis, we can see that the summation of the TFT scan time t_{TFT}, the LC response time t_{LC}, and the backlight flashing time t_{BL} must be shorter than the sub-field time, i.e., $1/(3f) \geq t_{TFT} + t_{LC} + t_{BL}$. For example, for a $60\,Hz$ image frame refreshing rate, $t_{TFT} + t_{LC} + t_{BL} \leq 5.56\,ms$. To yield sufficient brightness without dramatically increasing the number of LEDs, the backlight flashing time t_{BL} needs to be set at $\sim 1.5\,ms$ or even longer. If the LC response time t_{LC} is about $3\,ms$, then the total gate line scan should be completed within $1\,ms$. A short scan time budget limits the number of gate lines and imposes a certain amount of pressure on the present a-Si technology. In addition, from Figure 5.1, some pixels (e.g., those in the first row) reach the targeted saturation level much faster than the last ones and there is a waiting time before the backlight starts to flash for the whole panel. This waiting time increases as the number of gate lines scanned increases, which is a 'waste' in terms of utilizing the valuable time budget in each sub-field. To solve these problems, multi-area scanning schemes have been proposed [5, 8], as shown in Figure 5.2. In this multi-area scanning method, the whole panel is divided into N sub-blocks, each having an independent set of gate driver and local RGB LEDs to conduct color sequencing independently for each block. Thus, the number of gate lines for each gate driver can be decreased by N times, and the relation of the time division becomes $t_{TFT}/N + t_{LC} + t_{BL} \leq 1/(3f)$. Hence, the load on TFT scanning time is much reduced. Meanwhile, keeping the LC response time unchanged, the backlight flashing time is increased which improves the light output. Of course, the tradeoff is the increased number of gate drivers.

In addition, to make the output image more uniform, different blocks might not be exactly synchronized, instead there may be a short time delay from block to block in order to generate asynchronous backlight flashing but with some overlap, as Figure 5.3 shows [5]. Thus, when the local LEDs in block n are flashing (say R LEDs), there is a short overlapping period where LCs in blocks $n - 1$ and $n + 1$ have already reached the next gray level but LEDs there have

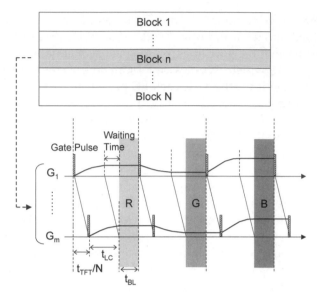

Figure 5.2 Illustration of multi-area driving scheme for a color sequential display

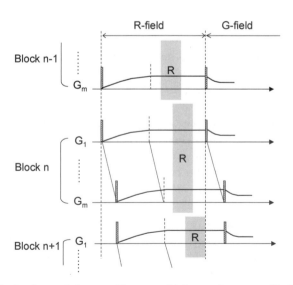

Figure 5.3 Illustration of the multi-area driving scheme with light coupling between adjacent blocks

not yet been turned on. Thus, the light from block n could be coupled into the adjacent blocks $n - 1$ and $n + 1$ even when they are still dark to produce a more uniform light output to the viewer. But the time sequence and light coupling intensity should be finely controlled to output the correct gray levels, which is quite complicated to implement practically. In this method, the time delay is small and the panel exhibits the same color for all blocks at a time. Yet in another multi-area scanning scheme, the delay between adjacent blocks is increased, thus different colors can flash from different blocks at the same time [8]. So when one block backlight flashes, there is leakage into adjacent blocks, and the delay time is designed to avoid this interference. In other words, within any single sub-field time (say, 5.56 ms), adjacent blocks are designed to appear either the same color, or one colored and the other black. For example, if we have three blocks from top to bottom when blocks 1 and 2 are both blue, block 3 must be dark; and when blocks 1 and 3 are red and blue, respectively, then block 2 must be dark. Meanwhile, in addition to multi-area scanning, for some large-panel applications, local color dimming, which divides the LEDs into different regions and controls the color light intensity from each region separately, can also be implemented to greatly enhance the contrast ratio and substantially reduce the power consumption.

5.3 Fast-response LC Modes

5.3.1 Thin Cells with High Birefringence LC Material

From the above analysis, a fast-response LC mode is critically important for color sequential LCDs. The most straightforward and efficient method of obtaining a fast response is to reduce the LC cell gap, since the response time is proportional to $\tau_o = \frac{\gamma \cdot d^2}{K \cdot \pi^2}$ [17, 18]. In turn, to gain sufficient phase retardation, a high birefringence (Δn) LC mixture is necessary. Presently, typical TFT-grade LC material has Δn around 0.1 and some commercially available TFT-grade LC mixtures have a high Δn around 0.2. For the application of color sequential, TN mode with high light transmittance and weak color dispersion is a good candidate. To reduce the color dispersion of a 90° TN cell, the required minimum $d\Delta n$ is about 480 nm at $\lambda = 550$ nm to meet the Gooch–Tarry 1st minimum condition [19]. Thus, the LC cell gap is about 2.4 μm for $\Delta n \sim 0.20$, leading to a response time around 5 ms with γ/K at about 10 ms/μm². From simulations using a TFT LC material with elastic constants $K_{11} = 9.7$ pN and $K_{33} = 13.3$ pN, dielectric anisotropy $\Delta\varepsilon = +8.2$, and rotational viscosity $\gamma_1 = 84$ mPa·s, the rise time and decay times of the 2.4 μm 90° TN cell are about 1.6 ms and 4.4 ms, respectively. To obtain a response time below 2 ms, a cell gap around 1.6 μm is necessary, so an LC material with $\Delta n \sim 0.30$ is required for light efficiency over 95%, or with $\Delta n \sim 0.20$ for reduced light efficiency at about 80%.

For high birefringence LC compounds, coupling a tolane or terphenyl rigid core and NCS terminal group would be a good candidate, owing to the reasonably high Δn, low viscosity, and good chemical, photo, and thermal stability [9]. The addition of lateral substitutions like F (fluorine) atoms to the tolane and terphenyl structures through LC molecular engineering could substantially lower the melting temperatures, making them suitable for display applications while keeping acceptable resistivity. An LC mixture with $\Delta n \sim 0.372$ at $\lambda = 632.8$ nm, $\Delta \varepsilon \sim 16.2$, and γ / K_{11} at about 12.5 ms/μm^2 has been developed based on the above concepts; its decay time in a 1.6 μm 90° TN cell is observed to be around 1.8 ms at room temperature [9], which is sufficient for color sequential applications. Although the Δn is high, the mixture is still transparent in the visible spectral region. In color sequential displays, RGB LEDs are employed. The blue wavelength (~450 nm) causes no harm to the LC material. As the wavelength decreases, say for green and blue LEDs, the LC birefringence increases. However, the voltage dependent transmittance curve of normally white TN cell is relatively insensitive to the wavelength because of the polarization rotation effect. Therefore, such a thin TN cell can be used for field sequential color displays. Considering the small panel size in mobile displays, mass fabrication of thin cells with a good yield is foreseeable.

5.3.2 Bend Cells

An OCB (optically compensated bend) cell [16] which utilizes a voltage-biased pi-cell [15] with compensation films is another candidate for fast-response and wide-view LCDs. Unlike conventional homogeneous cells, in a pi-cell, the surface pre-tilt angles are in a parallel direction, as shown in Figure 5.4. Below a certain voltage, called the *critical voltage*, the LC directors are more stable in a splay profile, which has the lowest energy. As the voltage increases above the critical voltage, certain nuclei will be generated near the surface and propagate towards the bulk region to transform the alignment into a low bend state. Under such circumstances, the bend profile exhibits a lower energy than the splay state. However, this transition from splay to bend takes from hundreds of milliseconds to even a few minutes before a stable bend state is achieved. There have been numerous successful efforts to develop methods that could speed up this initial 'splay to bend' transition and maintain the bend orientation even at a low voltage, such as using

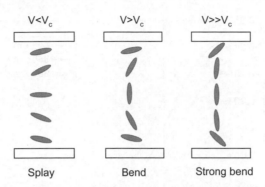

Figure 5.4 LC director configurations in a pi-cell at different voltages

polymer network formation [20–23], high surface pre-tilt angle [24, 25], or special driving schemes and fields with irregular electrode layouts [26, 27]. After this splay-to-bend transition is complete, the normal operation of the OCB cell can be performed in the bend state above the critical voltage. In comparison to other LC modes, the pi-cell or OCB cell exhibits a symmetrical wide viewing angle and rapid response time. The fast response originates from the following factors: the biased voltage effect, flow effect, and half-cell switching. The typical averaged gray-to-gray response time is below 3 ms, which is particularly interesting for LCD video applications. These properties make the OCB mode a strong contender for color sequential displays.

To investigate the electro-optics of OCB cells, let us numerically simulate a pi-cell by including the flow effect. The LC material is assumed to have $K_{11} = 9.3$ pN, $K_{33} = 11.7$ pN, and $\Delta\varepsilon = +15.0$; the Leslie coefficients accounting for the flow effect are assumed to be the same as the only available measured data from MBBA [28]. The surface pre-tilt angle is 9° and the cell gap is 4.2 μm. A uniaxial positive A-film with $d\Delta n \sim 57$ nm is placed with its slow axis perpendicular to the LC rubbing direction to compensate for the residual surface phase retardation of the OCB cell at 5 V_{rms}. This uniaxial A-film could also be replaced by a biaxial film to cancel the residual phase retardation and widen the viewing angle simultaneously. The calculated voltage-dependent transmittance (VT) curves at different wavelengths are plotted in Figure 5.5. Based on the material parameters, the critical voltage is about 1.43 V_{rms}. At $V = 1.5\,V_{rms}$, the light efficiency for R, G, and B light is 13.3% (out of the maximum possible value of 35.5%), 18.5% (out of the maximum possible value of 34.3%), and 23.9% (out of the maximum possible value of 28.6%), respectively. We can see that a low critical voltage is of particular importance in enhancing the transmittance of OCB cells. In

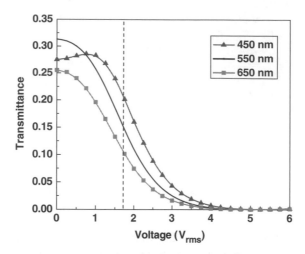

Figure 5.5 VT curves for an OCB cell at different wavelengths after compensation with a uniaxial A-film

addition to reducing the critical voltage for high transmittance, another alternative is to use a higher Δn LC material in the same cell gap. Thus, the transmittance peaks of R, G, and B wavelengths can be shifted towards the high-voltage side to get closer to or even above the critical voltage.

Widening the viewing angle for an OCB cell requires biaxial films or sophisticated discotic films in order to compensate for the HAN-cell-like surface director profile at both surfaces in the voltage-on state [29–33]. A typical optical configuration of a wide-view OCB cell is illustrated in Figure 5.6. Here, the rear and front polymer-discotic-LC (PDL) films function to compensate the boundary LC directors in the bend cell, and the residual phase retardation from the central LC cell is then compensated by an additional negative C-film placed between the PDL-film and the biaxial film. The films are placed on both sides of the pi-cell to enhance the viewing symmetry. Here, the rear and front biaxial films merely function to compensate the two crossed linear polarizers [34].

The detailed film parameters of the discotic films and negative C-films in the above wide-view configuration are related to the LC material, cell parameters, and operating voltage [31]. At a targeted voltage, say $5\,V_{rms}$, the combination of the negative C-films, the discotic films, and the pi-cell should output a zero or negligible overall phase retardation along the rubbing direction in all polar directions (including the axial direction where the C-film does not contribute to the polarization change). In other words, off-axis incident light traversing the stacks of the C-films, PDL-films, and the pi-cell

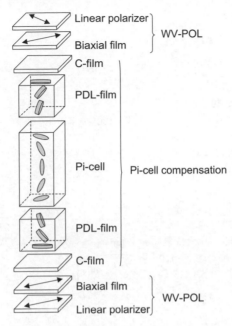

Figure 5.6 Optical configuration of film-compensated pi-cell

would roughly maintain its initial polarization. Hence, the two biaxial films could be employed to compensate only the effective angle deviation of the two crossed linear polarizers, as discussed in Chapter 3. At a selected dark state voltage, say $5\,V_{rms}$, the LC director profile in the pi-cell is known. Given the film birefringence data, the parameters remaining to be determined are: the rear and front inclination angles of the discotic film (the angle between each disc's optical axis and the substrate plane) θ_{rear} and θ_{front}, the discotic film thickness d_{disc}, and the negative C-film thickness d_c. Following this design concept, we find one set of optimal combinations: $\theta_{rear} \sim 42°$, $\theta_{front} \sim 64°$, $\Delta n \cdot d_{disc} \sim -75\,nm$, and $\Delta n \cdot d_c \sim -226\,nm$ with the dark state voltage at $V = 5\,V_{rms}$. Here, the LC rubbing direction is set at $-90°$ and the transmission axis of the rear linear polarizer is at $45°$. To illustrate the compensation mechanism, we simulated iso-contrast plots for three different compensation conditions at $\lambda = 550\,nm$, as shown in Figure 5.7. With only a uniaxial positive A-film, the viewing angle is quite narrow, as Figure 5.7(a) shows, indicating high off-axis light leakage. This single uniaxial film only functions to compensate the residual phase retardation from the pi-cell at normal incidence, but is insufficient for off-axis incidence resulting from the tilted LC directors. With discotic films and negative C-films, the viewing angle plot is shown in

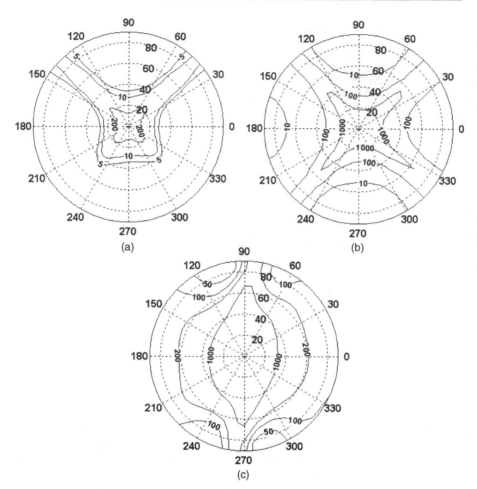

Figure 5.7 Iso-contrast plots for a pi-cell after compensation (a) with a uniaxial A-film, (b) with discotic films and negative C-films on both sides, and (c) with discotic films, negative C-films, and biaxial films

Figure 5.7(b). We can see that the viewing angle at ±90° is quite symmetrical and the 10:1 contrast ratio is maintained to greater than 55°. This is similar to an IPS cell between two linear polarizers, as discussed in Chapter 1. This indicates that the overall off-axis phase retardation from the three different layers is close to zero and the light leakage only arises from the two crossed linear polarizers. Furthermore, by using two additional biaxial films with $N_z = 0.25$ and 0.75 for the compensation of the two crossed linear polarizers, the viewing angle is dramatically widened, as shown in Figure 5.7(c).

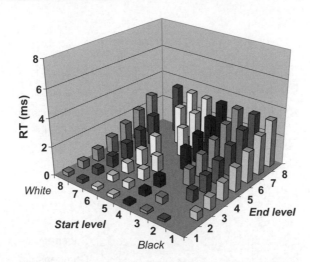

Figure 5.8 Gray-to-gray response time of an OCB cell

A contrast ratio $> 50:1$ could be achieved over 70° in all azimuthal directions, which is quite attractive for applications demanding a wide view.

For the above normally white OCB cell, the switching voltage is set between $1.5\,V_{rms}$ (white) and $5.0\,V_{rms}$ (black). Without using any overdrive method, the rise (switch-on) time from $V = 1.5\,V_{rms}$ to $V = 5.0\,V_{rms}$ is about 0.33 ms, and the decay time is about 3.34 ms. The gray-to-gray response times are plotted in Figure 5.8, where the average value is only about 1.6 ms. Here, the Leslie coefficients for the flow effect are assumed to be the same as those for MBBA. If the actual flow coefficients are taken into account for the material having the above-mentioned K, $\Delta\varepsilon$, and Δn values, the response time might be slightly different, but would still be in a similar range. By comparison, a typical 4-μm TN or VA cell has a response time (rise + decay) well above 25 ms without utilizing the overdrive and undershoot voltage method. Hence, the inherent fast response without any overdrive and wide viewing angle properties of bend cells are truly attractive for minimizing motion picture blur in LCDs.

5.4 Fast-response Transflective LCDs

5.4.1 Conventional Transflective LCDs Using OCB Modes

First we will briefly investigate conventional fast-response transflective OCB LCDs using color filters and divided transmissive and reflective regions.

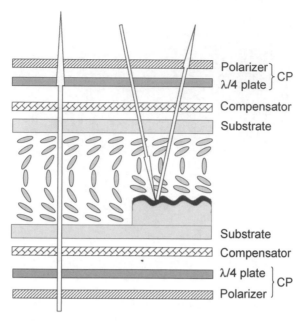

Figure 5.9 Device configuration of a transflective OCB LCD using dual cell gaps

A typical device configuration is shown in Figure 5.9, where different cell gaps in the transmissive region and reflective region are used to compensate for the optical path difference. A front circular polarizer comprised of a linear polarizer and a quarter-wave plate is placed in front of the reflective cell to obtain a dark state when the bulk LC directors are tilted up under a relatively high voltage. Here, the quarter-wave plate could be a monochromatic plate or a broadband one with a half-wave plate and a quarter-wave plate. Accordingly, the transmissive mode is sandwiched between two circular polarizers (CPs) to obtain a common dark state with the reflective mode. The two compensators function to compensate the residual axial phase retardation from the LC boundary directors in the dark state as well as to widen the viewing angle. A compensator could be a combination of a uniaxial A-film and a uniaxial C-film, a single biaxial film, or a combination of a discotic film and a biaxial film [35–42]. Here we will take one example using broadband circular polarizers and biaxial films as the compensators, as shown in Figure 5.10. Angle α depicts the transmission axis of the rear linear polarizer and angle β represents the rubbing direction of the pi-cell. In each circular polarizer, the optical axes of the half-wave plate and the quarter-wave plate

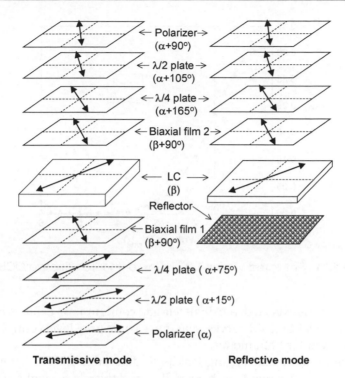

← Polarizer →
(α+90°)

← λ/2 plate →
(α+105°)

← λ/4 plate →
(α+165°)

← Biaxial film 2 →
(β+90°)

← LC →
(β)

Reflector

← Biaxial film 1
(β+90°)

← λ/4 plate (α+75°)

← λ/2 plate (α+15°)

← Polarizer (α)

Transmissive mode   **Reflective mode**

Figure 5.10 Optical configuration of a wide-view transflective OCB LCD using dual cell gaps and biaxial films as compensators

are set at 15° and 75° from the linear polarizer's transmission axis. The slow axes of the biaxial films are set perpendicular to the LC rubbing direction to cancel the on-state residual phase retardation.

To validate the device concept, a numerical simulation is conducted to compute the electro-optics of the device. The LC material used for the pi-cell is the same as that in the previous section, with $K_{11} = 9.3$ pN, $K_{33} = 11.7$ pN, and $\Delta\varepsilon = +15.0$. The pre-tilt angle on each surface is 9°. The LC cell gaps in the transmissive region and reflective region are chosen to be 4.2 μm and 2.1 μm, respectively. When determining the retardation values of the compensation biaxial film, the front biaxial film (Biaxial film 2) in the reflective sub-pixel is first taken into consideration, i.e., its in-plane phase retardation value is chosen so that the dark state of the VR curve occurs at 5 V_{rms}. Then, the rear biaxial film (Biaxial film 1) is arranged to make the VT curve dark at the same voltage. Following these guidelines, we found that $(n_x - n_y) \cdot d_{bx2} \sim 28.5$ nm and $(n_x - n_y) \cdot d_{bx1} \sim 28.5$ nm. The VT and VR curves at $\lambda = 550$ nm are plotted

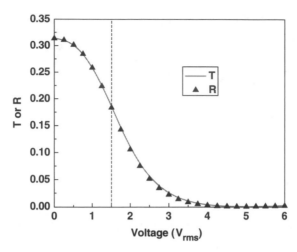

Figure 5.11 Simulated VT and VR curves for the transflective OCB LCD

in Figure 5.11. As expected, the dual-cell-gap configuration exhibits excellent overlap of the VT and VR curves, making a single gray level control gamma curve sufficient for both modes.

However, a more challenging task is to determine the N_z factors ($N_z = (n_x - n_z)/(n_x - n_y)$) of these two biaxial films and the related angles α and β of the rear linear polarizer and the LC rubbing direction that could yield a wide viewing angle. Balancing the reflective sub-pixel with the transmissive sub-pixel, one wide-view design using a Poincaré sphere is found to have $N_z \sim 5.5$ for both biaxial films, $\alpha = 25°$, and $\beta = -90°$. It is not necessary for these two biaxial films to be identical, but using the same biaxial film parameters simplifies the fabrication process. In practice, the biaxial film is laminated on to the circular polarizer during device fabrication. The iso-contrast plots of the transmissive and reflective sub-pixels using the above optical configuration and parameters are depicted in Figure 5.12. The viewing cone with CR > 10 : 1 is greater than 50° in most directions in both modes, which would be sufficient for small-panel mobile displays.

The above compensation using only biaxial films for the pi-cell would still generate weak gray-level inversion at certain viewing directions for the transmissive sub-pixel [39]. To eliminate this gray-level inversion, polymer discotic LC films (PDL-film) and biaxial film/C-film could be used for the compensators, as shown in Figure 5.13(a) [38–40]. However, owing to the inherent narrow-view property of crossed broadband circular polarizers, the improvement using PDL-films and C-films is still limited to a level similar

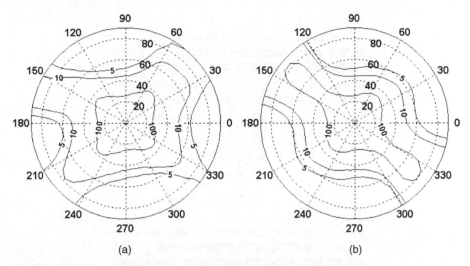

Figure 5.12 Iso-contrast plots of (a) the transmissive sub-pixel and (b) the reflective sub-pixel of a transflective OCB LCD using biaxial film compensation

to that shown in Figure 5.12. Therefore, in addition to the PDL-films and C-films, the use of wide-view broadband circular polarizers [43, 44] is necessary to widen the viewing angle.

In another optical compensation configuration, shown in Figure 5.13(b), the two PDL-films together with a negative C-film are placed in front of the LC cell and two wide-view circular polarizers are used [41, 42]. The rear PDL-film compensates the boundary LC director profile near the front substrate, the negative C-film cancels the central vertical LC directors, and the front PDL-film compensates for the surface LC directors on the rear substrate. In a wide-view circular polarizer configuration [33, 44], the viewing angle of the transmissive sub-pixel can be greatly expanded. However, considering the number of films involved (PDL-films and negative C-film), this configuration is very complicated and the display thickness increases dramatically. Better solutions to achieve wide-view transflective LCDs with fewer compensation films are technically challenging but practically important.

5.4.2 Color Sequential Transflective LCDs

Because it does not use color filters, a color sequential display is promising for tripling display brightness and resolution. To achieve sufficient luminance for outdoor application (over 500 cd/m^2 for most situations), the gain could be obtained in different ways. Boosting the backlight intensity by

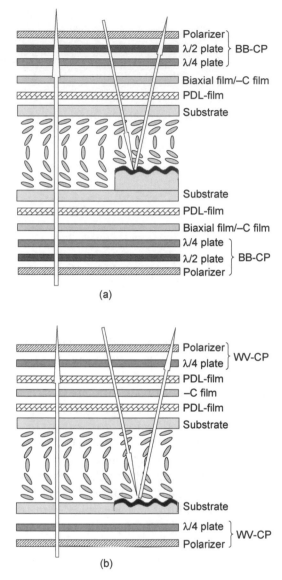

Figure 5.13 Device configuration of a transflective OCB LCD with discotic films (a) on both sides of the LC cell, and (b) only on the front of the LC cell

increasing the number of RGB LEDs would be a straightforward way. Using a faster response LC mode is also very important; therefore, high brightness could be obtained by increasing the backlight flashing time budget in each sub-field (~ 5.56 ms for a 60 Hz frame rate) by reducing the TFT scanning time and the LC response time. In addition, boosting the maximum attainable LC cell transmittance would also help. For color sequential based on the OCB mode, a low critical voltage pi-cell is very useful in order to raise the maximum possible transmittance of the LC cell itself. Besides focusing on backlight, for transflective LCD applications requiring outdoor readability, utilizing the reflectivity of some components could also be most helpful. Presently, there are several color sequential transflective configurations that are of interest [8, 45].

The first configuration, shown in Figure 5.14, is very simple and straightforward and uses the color-filter-free color sequential display for high reflectivity when the backlight is turned off [8]. A transflective sheet is laminated onto the rear linear polarizer surface facing the backlight. The transflective ratio between T and R can be adjusted according to different application requirements. This display has two major working modes: a color sequential transmissive display with backlight on and a monochromatic reflective display with backlight off. As shown in Figure 5.14, in the reflective mode, incident ambient light is first transmitted through the LC cell and the rear linear polarizer, then some of the light is reflected by the transflector, and the remaining portion goes further to the backlight unit and multiple reflections could occur. The portion that is reflected from the transflective sheet

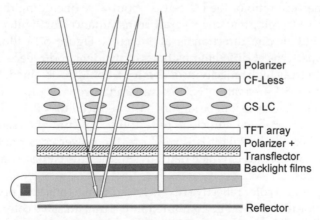

Figure 5.14 Transflective LCD configuration combining a color sequential transmissive display and a monochromatic reflective display

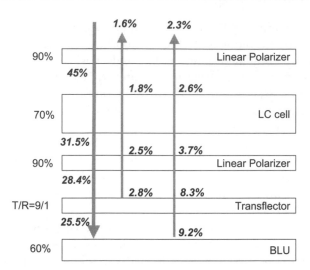

Figure 5.15 Reflectivity of incident and reflected beams at different locations

could maintain linear polarization. For light passing the transflector and the backlight films (including diffusers and BEF films), there will be multiple bounces with scattering from different elements and interfaces, thus the light polarization will be randomized. Overall, the reflectance could be high even if the transflector has a T/R ratio of about 9 : 1. To characterize the reflectivity of the display, we assume that each linear polarizer has a transmittance of about 90% for linearly polarized light that is parallel to its transmission axis, the effective aperture ratio of the LC cell is about 70% (including the transmittance of the LC cell), and the overall transmittance for light passing the backlight unit in one direction is about 60%. Figure 5.15 illustrates the reflectance patterns at different positions of incident and reflected light. We can see that a truly high reflectivity of about 4% could be obtained. In contrast, if color filters (33% area ratio × 85% of transmittance = 28% for a single pass) are used, the reflectance will be significantly reduced to ~ 0.3%. This simple concept would produce sufficient reflectivity for sunlight readability. In addition, when the backlight is turned off, the power consumption of the display is greatly reduced, which helps to lengthen the battery usage time for mobile devices.

To incorporate a color reflective display, traditional reflective pixels with R, G, and B color filters could be combined into a transmissive color sequential display [45], as shown in Figure 5.16. Each pixel in the device has four sub-pixels, three of which are for the reflective mode with bumpy reflectors and

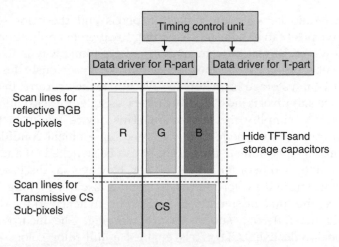

Figure 5.16 Device configuration of a transflective LCD combining a color sequential transmissive pixel and conventional reflective pixels with RGB color filters

different color filters and one of which is for the transmissive mode without a color filter. The three reflective sub-pixels share one scan line and the transmissive sub-pixel uses another separate scan line. The color sequential imaging method using switchable R, G, and B LEDs is applied to produce color images in the transmissive sub-pixel while ambient light with embedded color filters produces color images in the reflective sub-pixel. Each sub-pixel is driven by an individual TFT in the unit area. The driving TFTs and storage capacitors for all pixels can be buried under the reflective region to maximize the aperture ratio. Because of different driving schemes, different data drivers need to be used for the reflective sub-pixel and transmissive sub-pixel. In the reflective region, a displayed color is produced by the different gray levels in the R, G, and B sub-pixels. Once the voltage from the scan line turns on their TFTs, the data driver provides corresponding voltage levels to the sub-pixels and the image holds for the whole frame time (say, 16.67 ms for a 60 Hz refresh rate). For the color sequential transmissive sub-pixel, each R, G, and B sub-field is alternately applied within one-third of the frame time (say 5.56 ms for a 60 Hz refresh rate). In other words, the frequency of the scan driver and data driver are three times higher than those used in the reflective sub-pixel. In addition, the synchronization between the reflective data driver, the transmissive data driver, and the backlight flashing is critical to obtain good quality images.

The area ratio between the reflective pixels and the color sequential transmissive pixel can be adjusted according to different application requirements. However, in real implementation, an economic way is to form the reflective sub-pixel in front of those areas that were previously designed for driving TFTs and storage capacitors to fully utilize the aperture. In addition, because each sub-pixel is individually driven, a single-cell-gap configuration can be used to simplify the fabrication. This device works in the color sequential transmissive mode in low to medium ambient conditions. In a strong ambient environment, the backlight can be switched off and only the reflective mode used in order to save power. However, as discussed above, the reflectance from the color-filter-free region is also quite strong. To avoid interference, the transmissive pixels need to be switched to the dark state when only the reflective sub-pixels are working. The incorporation of conventional reflective LCD pixels enables a full-color color sequential display for mobile applications.

5.5 Summary

In this chapter we have presented color sequential technology for mobile displays. The driving schemes of color sequential displays were first reviewed. In a color sequential display, R, G, and B color fields are refreshed alternately at a rate three times faster than the image frame rate. In order to obtain sufficient light output, the backlight flashing time needs to be kept long enough in each sub-field. Thus, it requires a short gate line scanning time and a short LC response time. To reduce the gate line scanning time, besides upgrading the TFT material, such as by using poly-silicon instead of amorphous silicon, dividing the display panel into multiple blocks is also quite effective. Each block is scanned by an individual gate driver and local LEDs flash separately. To reduce LC response time, a thin TN cell using a high birefringence LC material and a bend cell were then investigated. For the bend cell, the electro-optical properties such as voltage-dependent transmittance and dynamic behavior were numerically investigated. In addition, the compensation scheme for achieving a wide viewing angle, such as using polymer discotic LC films, was also studied in detail. Finally, fast-response transflective LCDs were presented, such as conventional transflective OCB LCDs using color filters and color sequential transflective LCDs for high transmittance and reflectance. These fast-response displays, including color sequential displays, are particularly important for mobile displays.

References

[1] Armitage, D., Underwood, I. and Wu, S.T. (2006) *Introduction to Microdisplays*, John Wiley & Sons, Ltd, Chichester.

[2] Uchida, T., Saitoh, K., Miyashita, T. and Suzuki, M. (1997) Field-sequential full-color AMLCD without color filters. *IDRC '97*, pp. 37–40.

[3] Koma, N., Miyashita, T., Uchida, T. and Mitani, N. (2000) Color field sequential LCD using an OCB-TFT-LCD. *SID Tech. Digest*, **31**, 632–635.

[4] Koma, N. and Uchida, T. (2001) A novel display method for field sequential color LCD without color break-up. *SID Tech. Digest*, **32**, 400–403.

[5] Uchida, T., Ishinabe, T., Sekiya, K., Kishimoto, T., Seki, H., Kano, M., Ohizumi, M., Yamaguchi, M., Matsuda, A., Suzuki, Y., Kawashima, M., Taniguchi, Y., Sugiura, G., Uezono, S., Koyama, M., Kanazawa, Y., Suzuki, K., Tamura, K., Chiba, M., Okayama, T. and Yamamura, A. (2006) Color imaging and display system with field sequential OCB LCD. *SID Tech. Digest*, **37**, 166–169.

[6] Koma, N., Tanaka, Y., Mitsui, M. and Uchida, T. (2007) A fast-switching LCD for color-field-sequential projection displays. *SID Tech. Digest*, **38**, 991–994.

[7] Ishinabe, T., Miyashita, T., Uchida, T., Wako, K., Sekiya, K. and Kishimoto, T. (2007) High performance OCB-mode for field sequential color LCDs. *SID Tech. Digest*, **38**, 987–990.

[8] Tai, W.-C., Tsai, C.-C., Chiou, S.-J., Su, C.-P., Chen, H.-M., Liu, C.-L. and Mo, C.-N. (2008) Field sequential color LCD-TV using multi-area control algorithm. *SID Tech. Digest*, **39**, 1092–1095.

[9] Gauza, S., Zhu, X., Wu, S.T., Piecek, W. and Dąbrowski, R. (2007) High birefringence liquid crystals for color-sequential LCDs. *SID Tech. Digest*, **38**, 142–145.

[10] Arend, L., Lubin, J., Gille, J. and Larimer, J. (1994) Color breakup in sequentially scanned LCDs. *SID Tech. Digest*, **25**, 201–204.

[11] Baron, P.C., Monnier, P., Nagy, A.L., Post, D.L., Christianson, L., Eicher, J. and Ewart, R. (1996) Can Motion Compensation Eliminate Color Breakup of Moving Objects in Field-Sequential Color Displays. *SID Tech. Digest*, **27**, 843–846.

[12] Koma, N. and Uchida, T. (2003) A new field-sequential-color LCD without moving-object color break-up. *J. Soc. Info. Disp.*, **11**, 413.

[13] Sekiya, K., Miyashita, T. and Uchida, T. (2006) A simple and practical way to cope with color breakup on field sequential color LCDs. *SID Tech. Digest*, **37**, 1661–1664.

[14] Lin, F.-C., Huang, Y.-P., Wei, C.-M. and Shieh, H.-P. (2009) Color-breakup suppression and low-power consumption by using the Stencil-FSC method in field-sequential LCDs. *J. Soc. Info. Disp.*, **17**, 221.

[15] Bos, P.J. and Koehler-Beran, K.R. (1984) The π-cell, a fast liquid crystal optical switching device. *Mol. Cryst. Liq. Cryst.*, **113**, 329.

[16] Yamaguchi, Y., Miyashita, T. and Uchida, T. (1993) Wide-viewing-angle display mode for the active-matrix LCD using bend-alignment liquid-crystal cell. *SID Tech. Digest*, **24**, 273–276.

[17] Nie, X., Lu, R., Xianyu, H., Wu, T.X. and Wu, S.T. (2007) Anchoring energy and cell gap effects on liquid crystal response time. *J. Appl. Phys.*, **101**, 103110.

[18] Jiao, M., Ge, Z., Song, Q. and Wu, S.T. (2008) Alignment layers effects on thin liquid crystal cells. *Appl. Phys. Lett.*, **92**, 061102.

[19] Gooch, C.H. and Tarry, H.A. (1975) The optical properties of twisted nematic liquid crystal structures with twisted angles $\leq 90°$. *J. Phys. D.*, **8**, 1575–1584.

[20] Konno, T., Miyashita, T. and Uchida, T. (1995) OCB-Cell Using Polymer Stabilized Bend Alignment. ASID'95, pp. 581–583.

[21] Kikuchi, H., Yamamoto, H., Sato, H., Kawakita, M., Takizawa, K. and Fujikake, H. (2005) Bend-Mode Liquid Crystal Cells Stabilized by Aligned Polymer Walls. *Jpn. J. Appl. Phys.*, **44**, 981.

[22] Hasegawa, R., Kidzu, Y., Amemiya, I., Uchikoga, S. and Wakemoto, H. (2007) Electro-optical properties of polymer-stabilized OCB and its application to TFT-LCD. *SID Tech. Digest*, **38**, 995–998.

[23] Yang, B.-R., Elston, S.J., Raynes, P. and Shieh, H..-P. (2008) High-brightness relaxed-bend state in a pi cell stabilized by synchronized polymerization. *Appl. Phys. Lett.*, **92**, 221109.

[24] Xu, M., Yang, D.K., Bos, P.J. and Jin, X. (1998) Very high pretilt alignment and its application in pi-cell LCDs. *SID Tech. Digest*, **29**, 139–142.

[25] Kim, J.B., Kim, K.C., Ahn, H.J., Hwang, B.H., Kim, J.T., Jo, S.J., Kim, C.S., Baik, H. K., Choi, C.J., Jo, M.K., Kim, Y.S., Park, J.S. and Kang, D. (2007) No bias pi cell using a dual alignment layer with an intermediate pretilt angle. *Appl. Phys. Lett.*, **91**, 023507.

[26] Nakao, K., Suzuki, D., Kojima, T., Tsukane, M. and Wakemoto, H. (2004) High-speed bend transition method using electrical twist field in OCB mode TFT-LCDs. *SID Tech. Digest*, **35**, 1416–1419.

[27] Wu, C..-H., Shih, P..-S., Yang, J.H., Chen, P..-Y., Chang, J., Pan, H..-L., Hsu, K.F., Lin, C..-Y., Lin, S..-H., Chang, C..-C. and Yang, K..-H. (2007) Development of 3.5″ VGA field-sequential-color optically compensated birefringence a-Si: H TFT LCDs with fast splay-to-bend transition and high cell transmittance. IDW'07, pp. 33–36.

[28] Wang, H., Wu, T.X., Gauza, S., Wu, J.R. and Wu, S.T. (2006) A method to estimate the Leslie coefficients of liquid crystals based on MBBA data. *Liq. Cryst.*, **33**, 91–98.

[29] Mori, H. and Bos, P.J. (1999) Optical performance of the π cell compensated with a negative-birefringence film and an A-plate. *Jpn. J. Appl. Phys.*, **38**, 2837.

[30] Ito, Y., Matsubara, R., Nakamura, R., Nagai, M., Nakamura, S., Mori, H. and Mihayashi, K. (2005) OCB-WV film for fast-response-time and wide-viewing-angle LCD-TVs. *SID Tech. Digest*, **36**, 986–989.

[31] Fukuda, I., Nakata, T., Sakamoto, Y., Ishinabe, T. and Uchida, T. (2006) Optimization of viewing-angle optical properties in an OCB-LCD compensated with hybrid-aligned discotic liquid-crystal films and c-plates. IDW'06, 133–136.

[32] Ishinabe, T., Ohno, Y., Miyashita, T., Uchida, T., Yaginuma, H. and Wako, K. (2006) Development of super high performance OCB mode for high quality color-field sequential LCDs. *SID Tech. Digest*, **37**, 717–720.

[33] Ishinabe, T., Miyashita, T., Uchida, T., Wako, K., Kishimoto, T. and Sekiya, K. (2004) Improvement of transmittance and viewing angle of the OCB mode LCD by using wide-viewing-angle circular polarizer. *SID Tech. Digest*, **35**, 638–641.

[34] Ishinabe, T., Miyashita, T. and Uchida, T. (2002) Wide-viewing-angle polarizer with a large wavelength range. *Jpn. J. Appl. Phys.*, part 1, **41**, 4553–4558.

[35] Lee, S.-R., Jung, M.J., Park, K..-H., Yoon, T..-H. and Kim, J..-C. (2005) Design of a transflective LCD in the OCB mode. *SID Tech. Digest*, **36**, 734–737.

[36] Chang, T..-J. and Chen, P..-L. (2003) A novel optically compensative structure for gray-scale-inversionless OCB-LCDs. IDW'03, pp. 77–80.

[37] Yao, I..-A., Yang, C..-L., Chen, C..-J. and Pang, J..-P. (2006) Optical design of wide-viewing-angle transflective OCB LCD. IDW'06, pp. 145–148.

[38] Yao, I..-A., Ke, H..-L., Yang, C..-L., Chen, C..-J., Pang, J..-P., Chen, T..-J. and Wu, J..-J. (2006) Electrooptics of transflective displays with optically compensated bend mode. *Jpn. J. Appl. Phys.*, **45**, 7831–7836.

[39] Xiang, R..-J., Yang, C..-L., Chen, C..-J. and Chen, S..-H. (2005) Electro-optical optimization in transflective OCB LCD. IDMC'05, pp. 597–600.

[40] Yao, I..-A., Chang, S..-H., Chen, C..-J., Yang, C..-L. and Pang, J..-P. (2008) A novel transflective OCB LCD for mobile TV applications. *SID Tech. Digest*, **39**, 495–498.

[41] Fukuda, I., Ohnishi, T., Ishinabe, T. and Uchida, T. (2006) A new transflective OCB-LCD with fast response time and wide-viewing angle. IDW'07, pp. 1543–1546.

[42] Fukuda, I., Ohnishi, T., Ishinabe, T. and Uchida, T. (2008) A new single-cell-gap transflective OCB-LCD with fast response time and wide viewing angle. *SID Tech. Digest*, **39**, 499–502.

[43] Ge, Z., Jiao, M., Lu, R., Wu, T.X., Wu, S.T., Li, W.Y. and Wei, C.K. (2008) Wide-view and broadband circular polarizers for transflective liquid crystal displays. *J. Disp. Technol.*, **4**, 129–138.

[44] Ishinabe, T., Miyashita, T. and Uchida, T. (2001) Design of a quarter wave plate with wide viewing angle and wide wavelength range for high quality reflective LCDs. *SID Tech. Digest*, **32**, 906–909.

[45] Lee, J.H., Zhu, X. and Wu, S.T. (2007) Novel color-sequential transflective liquid crystal displays. *J. Disp. Technol.*, **3**, 2–8.

6

Technological Perspective

6.1 Unique Role of Transflective LCDs

The rapidly expanding market for smart mobile display devices is driving a need to develop good image quality (high brightness, high contrast ratio, wide viewing angle, and sunlight readability for information display) and multi-functions (touch panels for a user-friendly interface and new applications). We will look at the technological perspectives of transflective LCDs by discussing their unique capabilities and technological advances to meet these two major requirements. In comparison with a pure transmissive display, a transflective LCD is a compromise between transmissive and reflective modes. To obtain a similar high indoor image quality to a transmissive display, some issues, such as a relatively low contrast ratio and a narrow viewing angle, remain unsolved or only partially solved. Meanwhile, the reflectivity of a transflective LCD is not as high as that of a pure reflective LCD. In addition, when new functions like touch panels are incorporated into mobile transflective LCDs, the optical architecture becomes much more complicated and some new issues emerge. These are some major technical challenges that the transflective LCD industry needs to address in the near future.

In this section we will first use one example to illustrate the importance of transflective functions for sunlight-legible LCDs and describe in detail the future development directions from the LC viewpoint to enhance the

Transflective Liquid Crystal Displays   Zhibing Ge and Shin-Tson Wu
© 2010 John Wiley & Sons, Ltd

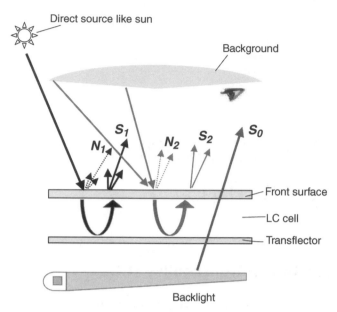

Figure 6.1 Light illumination pattern of a transflective LCD device

performance of transflective LCDs. This could also serve as a rough gauge for researchers and engineers to select the proper display technology from various choices like transmissive LCDs, organic LEDs, reflective LCDs, E-paper, and transflective LCDs, when sunlight readability is considered an important issue for the display.

In Figure 6.1 we have redrawn the illumination pattern of a typical display in a sunlit environment from Chapter 1. In a real outdoor environment, there are multiple light sources such as external direct sunlight, internal backlight, and much background light accounting for reflected or scattered sunlight from different objects. The direct sunlight may be viewed as a collimated light source, and the background light comes from all directions behaving like a broad area light source having a hemispherical, dome-like profile. The illuminance level of direct sunlight can reach over 100 000 lux and the typical background light illuminance level ranges from 10 000 to 25 000 lux and the maximum value could reach over 40 000 lux [1, 2]. The front surface of a display can be grouped into three categories: (i) smooth specular surfaces, (ii) surfaces with anti-glare (AG) coatings, and (iii) specular surfaces with anti-reflection (AR) coatings, where the AG surface is a combination of diffusive, spreading, and specular reflections [2]. The reflection patterns are highly dependent on the light source and front surface feature. For example, for direct sunlight illumination on a smooth specular polarizer surface, the reflected light will be highly collimated in the specular direction but still with a small amount of reflection away from that centered direction; and the reflected light from direct sunlight illumination upon an AG surface will be scattered to a broad range of angles with combinations of diffusions and specular reflections in accordance with the haze level of the AG surface. For the background light source, the reflected light can be viewed as uniformly distributed across all angles in the top hemisphere. Typically, a specular surface with AR coatings will be ideal for high contrast ratio and image legibility in real mobile displays for outdoor applications.

Because there is no strict standard for evaluating image readability of a transflective display in a real ambient situation, we propose a quantitative estimation method [3]. First, we introduce the concept of an ambient contrast ratio (ACR) in accordance with Figure 6.1. S_0, S_1 and S_2 shown in the figure represent the received light luminance in the bright state from the LC backlight, from direct sunlight after traversing and reflection from the LC cell, and from ambient background light after traversing and reflection from the LC cell, respectively. The received light in the bright state is expressed as $(S_0 + S_1 + S_2) + (N_1 + N_2)$, while the light received in the dark state is $(N_1 + N_2) + (S_0)/CR_0 + (S_1 + S_2)/CR_1$. Hence, the ambient display contrast ratio is $[(S_0 + S_1 + S_2) + (N_1 + N_2)]/[(N_1 + N_2) + (S_0)/CR_0 + (S_1 + S_2)/CR_1]$, where CR_0 is the LCD transmissive contrast ratio measured in a dark room, showing its capability to block the backlight (S_0) in the dark state, and CR_1 is the contrast ratio of the reflective sub-pixel of the display viewed under ambient light illumination (S_1 and S_2) (neglecting surface reflection). CR_0 and CR_1 are typically over 100 and make $(S_0)/CR_0 + (S_1 + S_2)/CR_1$ negligible. Hence, based on the contrast ratio definition that is taken as the ratio between light received in the bright state and that received in the dark state, the ambient contrast ratio can be written as $ACR \approx [(S_0 + S_1 + S_2) + (N_1 + N_2)]/(N_1 + N_2) \approx (S_0 + S_1 + S_2)/(N_1 + N_2) + 1$. In this equation we have neglected noise including: (i) the ambient light directly impinging on the human eye without interacting with the display panel (after considering the real human viewing pattern), and (ii) the intrinsic light leakage from the LCD itself (as the intrinsic contrast ratio of the LCD is usually well over 100:1). To estimate the sunlight readability of a display, Table 6.1 summarizes the readabilities of a display with different ACR values. As a reference, a printed paper has a typical ACR of about 15:1, which looks as good as we can experience in our daily lives. And for a typical notebook computer with a $200 \, \text{cd}/\text{m}^2$ surface luminance and a front surface reflectivity

Table 6.1 Sunlight readability according to different ambient contrast ratios (ACR) (3)

ACR	Outdoor Readability
<2	Unreadable
3–4	Readable in shade, barely readable in sunlight
5–9	Adequately readable in sunlight
10–14	Very readable in sunlight and looks good
>15	Excellent readability and looks great

of 2%, the ACR is usually less than 2 under direct sunlight. Thus, the image is unreadable or barely readable, which also agrees with our common understanding.

From the above information, we can compare the sunlight legibility of a transmissive LCD, a transmissive OLED, and a transflective LCD. This enables us to calculate display specifications that would lead to good image legibility. Let us assume that direct sunlight has an illuminance of 100 000 lux and that background light (clear sky) reaches 25 000 lux for a typical sunny day. When incident light is reflected, the luminance of the reflection will be measured as the luminous flux per unit solid angle with a unit of candela per square meter (cd/m^2). To convert illuminance to luminance, for an ideal diffusive surface with 100% reflectivity, the illuminance needs to be divided by a factor of π. Thus, a 100% diffusive reflective surface would have luminance of $\sim$31 800 cd/m^2 from direct sunlight and $\sim$7960 cd/m^2 from background light. If we assume the front surface is a typical specular surface with an AR coating having 1% reflectivity, the specular reflection of direct sunlight would be $\sim$318 cd/m^2, which would definitely wash out a 200 cd/m^2 backlit notebook screen viewed from the specular reflection direction. In practice, direct sunlight impinges upon the panel surface at an angle from about 20° to 40° from the display normal, and the viewer views the image at the normal direction, so avoiding the strong direct specular reflection. According to measured data [2], the reflectivity of a display surface at the normal direction of light incident from 25° and 35° ranges from 0.2% to 0.3%. (This value would vary with different AR-coated polarizer surfaces.) This results in $N_1 \sim 80\,cd/m^2$ (using a reflectivity value of 0.25%) in the normal direction. On the other hand, the background light also contributes a reflection $N_2 \sim 80\,cd/m^2$ (1% AR surface reflection) in all directions. Thus, the total reflection from these two sources is about 160 cd/m^2 when the viewer observes images from a direction normal to the panel surface. Interestingly, the contribution of noise from direct sunlight and the background is about the same magnitude in this example, where the image is not viewed from the specular reflection direction.

From Table 6.1, to obtain $ACR \sim 4$ for a transmissive display to be readable in sunlight, we could take the following approaches: (i) use a lower reflectivity AR coating on the display surface, (ii) view the display at roughly normal direction to avoid surface specular reflections, and (iii) implement adaptive backlight brightness control, which means the backlight brightness increases in proportion to the ambient. Under direct sunlight, the backlight should provide a panel surface luminance of $\sim$500 cd/m^2. There are several problems associated with these approaches. First, boosting the backlight brightness would significantly increase the power consumption, reducing the battery life and generating a lot of heat. Second, a high brightness of the display would decrease the dark pixel´s ability to block light, making the dark color appear gray. Therefore, developing multilayered AR coatings with reflection down to 0.5% would be a good solution. Under such circumstances, a 200 cd/m^2 surface brightness leads to an $ACR \sim 3.5$ ($ACR = (S_0 + S_1 + S_2)/(N_1 + N_2) + 1$, where $S_1 = S_2 \sim 0$ for pure transmissive LCDs, and $N_1 = 31\,800 \times 0.25\% \times 1/2 \sim 40$ cd/m^2, and $N_2 = 7960 \times 0.5\% \sim 40$ cd/m^2) But for mobile devices like touch-panel cell phones, frequent touching of the front surface by human fingers would gradually degrade the AR coating performance and the cost for multilayer AR coating is quite expensive. Nevertheless, for a pure transmissive display to maintain good image readability in strong sunlight, both adaptive backlight brightness control and robust low-reflectivity AR coatings need to be implemented.

On the other hand, let us compare the ACR of a transflective LCD with surface luminance at 200 cd/m^2 by the backlight and a normal 2% reflectivity from the reflective pixels. The front surface is also a specular surface with AR coatings having 1% front surface reflectivity. Under a similar illumination condition, $N_1 + N_2$ is about 160 cd/m^2, $S_0 = 200$ cd/m^2, and $S_2 = 7960 \times 2\% \sim 160$ cd/m^2. However, the major difference here comes from the S_1 part, which accounts for the useful reflection from the reflective LCD pixels of the direct sunlight. Using a diffusive bumpy reflector or even a diffusive micro-slant bumpy reflector as discussed in Chapter 1, the 30° incident direct sunlight impinging on the bumpy reflector can be deflected towards the normal direction of the viewer. Thus, S_1 (from 100 000 lux direct sunlight) will be significantly greater than S_2 and up to several hundred cd/m^2. In brief, the $ACR = (S_0 + S_1 + S_2)/(N_1 + N_2) + 1$ could easily reach 5 to 6, making the image adequately readable even in strong sunlight. In addition, as the ambient light intensity increases (S_1 and S_2 become much larger than S_0) the ACR roughly approaches the inherent value coming from the reflective display sub-pixel, which is $(S_1 + S_2)/(N_1 + N_2) + 1$ or $ACR \sim 4$ to 5, independent of the backlight. Hence, using a transflective LCD, the ACR value is less dependent on the backlight brightness under strong ambient conditions, and the backlight could even be turned off to save power. Similarly, by suppressing the AR coating reflectivity to 0.5%, the ACR would be improved to over 10 to give a well readable image quality.

This analysis gives us a clear pattern of how ambient conditions affect the image legibility and which display technology (among transmissive LCDs, OLEDs, reflective LCDs, E-paper, and transflective LCDs) should be chosen for special applications. It can also clearly be shown that transflective LCDs with both transmissive and reflective modes exhibit unique characteristics for mobile applications. Nevertheless, the major tradeoff of transflective LCDs is the compromise between transmissive and reflective modes. For example, in MVA-based mobile transflective LCDs, circular polarizers are used to obtain a normally black reflective mode and transmissive mode simultaneously. With circular polarizers, the image quality, in terms of contrast ratio and viewing angle, could be greatly degraded for the transmissive sub-pixel in comparison with that obtained using crossed linear polarizers. In IPS- or FFS-based transflective LCDs, the transmissive sub-pixel could be simply placed between crossed linear polarizers, but the reflective mode is quite troublesome; it demands a complex treatment such as forming in-cell retarders in order to secure a normally black mode. Therefore, designing novel wide-view circular polarizers for mobile MVA transflective LCDs, and developing novel IPS or FFS transflective LCDs with easy fabrication and high optical performance should be the central research focus and development target. In addition, for the reflective sub-pixel, both high contrast ratio and high brightness are almost equally important for human eyes to perceive high-quality images. Designing high-reflectivity bumpy reflectors and increasing the efficiency of the reflective pixels are of particular importance to enhancing the image quality of transflective LCDs.

6.2 Emerging Touch Panel Technology

Next, we will further discuss some key design considerations when combining touch panel technology with mobile transflective LCDs and potential technological advances for high-performance touch panel mobile LCDs.

Touch panel has been convincingly advancing to next generation display technology. Therefore, technological elucidation about touch panel mobile displays is of particular interest and importance. In the present market, there are various touch technologies targeted at both small-panel and large-panel applications [4]. For touch panel AMLCDs, the technology may be categorized as the external type, which stacks a touch screen in front of the LCD cell, and the internal type, which embeds the touch panel function into the LC cell.

For external-type touch panels, one of the popular technologies is the low-cost resistive touch screen aimed at low-end applications. The device configuration is depicted in Figure 6.2. From the left-hand diagram, two conductive transparent ITO planes are separated with an air gap by the spacers on the rear substrate, and a front flexible protective membrane like PET is coated on the front surface. When a finger or a stylus presses the front surface, the two ITO planes electrically touch each other at that point, which can be detected by the embedded sensing electronics. The right-hand diagram in Figure 6.2 shows one example, the four-wire resistive touch screen, to illustrate the method for sensing the location of the touch point. First, to determine the touch point coordinate in the horizontal direction, voltage is applied between two side strip electrodes in the rear ITO plane, generating a potential gradient in the horizontal direction. The front ITO plane functions as a sensing layer to measure the voltage value of the touching point and to map out the horizontal coordinate. Voltage is then applied across the vertical direction on the front substrate, and the rear layer works as the sensing plate to find out the vertical direction. In other designs, such as five-wire or six-wire resistive touch screens, only the front plate functions as the sensing plate and the voltages are applied alternately across the horizontal and vertical directions on the same rear substrate. The resistive-type touch screen is relatively simple, cheap, and works with many objects like finger and stylus. It also exhibits a relatively low optical transmittance (10% to 20% light loss) and cannot support multi-touch. This technology is widely applied in touch panel mobile devices where cost is the most important issue.

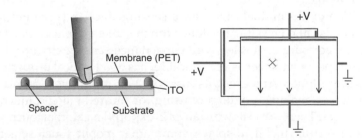

Figure 6.2 Illustration of a resistive touch screen

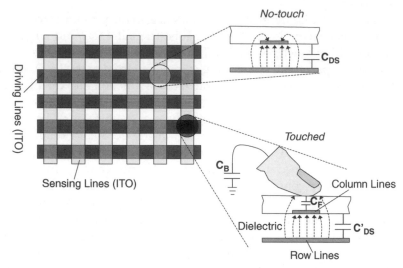

Figure 6.3 Illustration of a mutual capacitance-based touch screen

Another widely adopted external touch panel technology is the capacitive-type touch screen. Figure 6.3 depicts a capacitive touch screen based on the mutual capacitance change between driving lines and sensing lines. When a pixel is not touched (as in the top-right diagram), the mutual capacitance C_{DS} between driving and sensing lines there remains a constant. On the other hand, when an object like a finger approaches or touches one pixel region (as in the bottom-right diagram), the capacitance of the finger C_F and the human body capacitance C_B will be automatically coupled to the mutual capacitance between driving and sensing lines, and will change it to C'_{DS}. When a scanning signal is applied row by row from front to rear, the sensing lines can sense a change of current (or voltage) at the specific touched pixel to determine its coordinate. With this method, multiple-touch is also attainable.

In addition to this mutual capacitance sensing, another type of capacitive touch screen is called a self-capacitance touch screen; this uses one layer of individual electrodes connected with capacitance sensing circuitry [5]. The capacitive-type touch screen is very durable, has good optical properties, and supports multi-touch. At present, its cost is still high, but this is decreasing as usage increases, making it quite promising for future high-end touch panel mobile devices. These are just two examples of popular external touch screens widely adopted for mobile displays; some other mobile touch screen technologies like surface force sensing touch screens are also emerging [4].

Let us now discuss some issues related to combining external touch screens with a typical LCD panel. When stacking an external touch screen with an LC cell, there will be many more interfaces that could generate undesired reflections due to refractive index mismatch. For example, in the conventional resistive touch screen, between the front and rear ITO planes, there is an air gap. If no AR coating is deposited on the ITO surface, the reflection from a single ITO–air interface will be well over 5% due to the high refractive index of ITO. This would easily kill the outdoor image readability of the touch panel incorporated LCD, even with $500\,cd/m^2$ brightness. Utilizing multilayer AR coating would partially overcome this problem, but the cost would be increased dramatically.

A good alternative solution is to use a circular polarizer to suppress the internal reflections in touch screen incorporated LCDs. Figure 6.4 illustrates the cross-sectional view of an LCD using a circular polarizer in front of the front touch screen. Here, the front surface of the hardcoat is coated with an AR-film with 0.5% reflectivity. A high-quality circular polarizer could suppress the internal reflection at each ITO–air interface of the touch cell unit down to 0.1%. Please note that the membrane in front of the front ITO in the touch unit must have zero or negligible retardation to make the circular polarizer functional in blocking the internal reflections. With regard to the assembly of the touch screen and the LC cell, there might also be other interfaces with mismatched refractive index, such as that between a glass substrate in the front touch cell and the retardation film belonging to the rear LC cell unit. But the internal reflections could also be blocked by the front circular polarizer. As a result, the total reflection would be suppressed below 1%. From the analysis in the previous section, a transmissive display at a

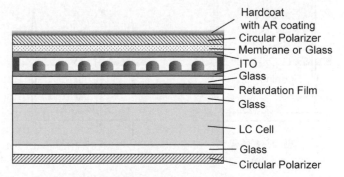

Figure 6.4 Illustration of resistive touch screen using circular polarizers and AR coatings to suppress the reflections

surface brightness of $500\,\mathrm{cd/m^2}$ from backlight could yield an $ACR\,(=500/\,160+1)\sim 4$ to get a readable outdoor image in strong sunlight; and a transflective LCD with a $200\,\mathrm{cd/m^2}$ backlight illuminated brightness and 2% reflectivity from the reflective pixel would reach an $ACR \sim 5\text{--}6{:}1$, which is adequate for sunlight readability. To further suppress the internal reflections, the ITO surfaces need to be treated well in both resistive and capacitive touch screens.

From the above analysis, stacking an external touch screen with an LCD would be the most straightforward method for touch panel mobile displays, but the drawbacks are increased thickness and multiple internal reflections that could reduce the image quality. Therefore, embedding the touch function into the LC cell would be highly desirable. Below, we will briefly introduce some typical in-cell touch panel technologies.

A very simple in-cell touch panel technology is based on optical sensors [6–8], as illustrated in Figure 6.5. Within each single pixel or a few pixels, an optical sensor along with related sensing electronics is embedded on the rear TFT substrate. Under strong ambient light, such as outdoor conditions, each photo sensor can detect the incident ambient light intensity, outputting a light intensity map to the processor to determine if the area is shielded by an object like a finger, as shown in Figure 6.5(a). When the ambient conditions are quite dim, as in a dark room, the device can shift to work with a reflective mode to detect the reflected backlight by the touch object towards the photo sensor, as shown in Figure 6.5(b). With this technology, relatively good touch sensitivity

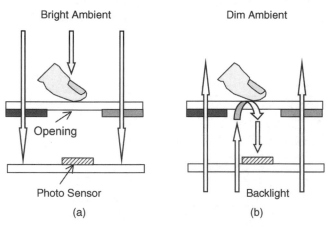

Figure 6.5 Illustration of an in-cell touch screen based on photo sensors under (a) strong ambient conditions and (b) low ambient conditions

can be obtained in this mode under bright ambient conditions, but the touch sensitivity in low ambient usage is low and might be a big problem; for example, touching a black pixel could result in no response. In addition, the aperture ratio of the device may also be reduced, as the optical sensor occupies a separate region that blocks light transmission.

The in-cell capacitive-type touch screen is another popular technology, which detects the change in LC capacitance to determine whether the pixel region is being touched or not [7–11]. The device configuration is shown in Figure 6.6(a), where a mutual capacitance is formed between the rear sensing electrode and the front conductive protrusion surface. The capacitance value is proportional to the LC effective dielectric permittivity and the gap distance. When a pixel is not touched, all local LCs are uniformly distributed as in a vertical alignment and the gap is a constant. Under touch pressure, the local LC directors are deformed (the effective permittivity changes) and the gap shrinks, resulting in a change in the mutual capacitance, which can be detected by the sensing electrode and related electronic circuit. Unlike the above light-sensing method, this technology is independent of the ambient conditions and light source, and also has a higher aperture ratio. In addition, the fabrication of the capacitance-sensing unit is compatible with the present fabrication process, resulting in less change in the LCD backplane compared with the light-sensing approach. In some devices, capacitive-type touch sensing is combined with the light-sensing method to gain sufficient touch sensitivity [7, 8]. In another emerging method, called in-cell resistance sensing (depicted in Figure 6.6(b)), the touch signal is generated by direct mechanical contact between two conductive surfaces from the sub-column spacer and the rear sensing electrode [12–14]. Unlike the capacitance method, the gap between the column spacer conductive surface and the rear sensing electrode is very small (even down to ~0.5 µm). When the pixel region is touched, the

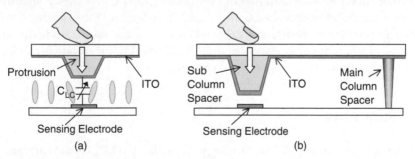

Figure 6.6 Illustration of an in-cell touch screen based on (a) LC capacitance sensing and (b) mechanical touch sensing

column spacer surface approaches and finally touches the rear electrode, and signals can be detected by the backplane sensing electronics. Similar to the in-cell capacitive-sensing method, this technology provides clear images with high light transmission and fast speed with integrated sensing and processing circuitry. But LCD surface softness and surface pooling is a big problem for this technology that requires good solutions. Nevertheless, compared with external touch screens, all these in-cell touch technologies support multi-touch, have much reduced thickness and weight, output clearer images, and give users truer touch experience, making them potential candidates for future touch panel technologies.

One important practical issue for LCs related to touch panels is the restoring time of LC directors after the touch pressure is released. Previously, vertical alignment-based LC modes were not preferred for touch panel application, because of the long restoring time associated with the press–release process. For vertically aligned LC directors, once the surface pressure deforms the substrates, the LC directors undergo a complicated reorientation, involving both vertical tilt down and in-plane twist. In the restoring stage, irregular movements take a long time for the LCs to return to their initial vertical positions. On the other hand, homogeneous alignment using positive LC material like IPS or FFS cells is more robust for application to touch panel screens. Fortunately, for a VA LC cell, the recently developed polymer-sustained surface alignment technology is an effective method of dramatically shortening the restoring time for touch panel applications. This method is still under active research [15–17]. For IPS or FFS cells, the rear and front alignment surfaces are mechanically rubbed to have strong anchoring forces for fixing the boundary LC directors. When a voltage is applied, the bulk LC directors are reoriented mainly in the horizontal direction by electric fields. Hence, even after some deformation by touch pressure, the change in LC director distribution is much weaker than that in vertical alignment, and strong anchoring forces can quickly restore the LC alignment. To speed up the restoring time or make the panel more resistant to external touches, the pixel electrode design of IPS and FFS is quite important [18, 19]. Presently, IPS/FFS mode is attracting more attention for touch panel applications.

6.3 Summary

Similar to mainstream LCD technology, transflective LCD is also advancing rapidly and efficiently in almost every component level, including LC material, the backlight unit, and driving circuits. With good indoor and outdoor image readability, transflective LCDs have intrinsic advantages over transmissive LCDs. For example, in some emerging applications like touch panel LCDs using external touch screens, there are always some residual reflections from multiple interfaces; a transflective LCD with circular polarizers provides good readability by the reflective sub-pixel without dramatically increasing the backlight power. Transflective LCDs will continue to play irreplaceable roles in the mobile display market. The ultimate goal is to have the transmissive sub-pixel comparable to pure transmissive LCDs under linear polarizers and have the reflective sub-pixel with high reflectivity, high contrast, and good color saturation, which could be systematically attained from novel LCD cell configuration development, compensation film design, and material engineering.

References

[1] http://en.wikipedia.org/wiki/Daylight.
[2] Chung, H.-H. and Lu, S. (2003) Contrast-ratio analysis of sunlight-readable color LCDs for outdoor applications. *J. Soc. Info. Disp.*, **11**, 237.
[3] Walker, G. (2007) GD-Itronix DynaVue Display Technology – The Ultimate Outdoor-Readable Touch-Screen Display. Available at: www.ruggedpcreview. com.
[4] Walker, G. (2009) *Emerging touch technologies*. SID short course, San Antonio, Texas, June 1.
[5] Hotelling, S.P., Chen, W., Krah, C.H., Elias, J.G., Yao, W.H., Zhong, J., Hodge, A.B., Land, B.R. and Boer, W.D. (2008) Touch screen liquid crystal display. U.S. Patent US2008/0062139 A1, March 13.
[6] den Boer, W., Abileah, A., Green, P., Larsson, T., Robinson, S. and Nguyen, T. (2003) Active matrix LCD with integrated optical touch screen. *SID Tech. Digest*, **34**, 1494–1497.
[7] Kang, M.-K., Uh, K. and Kim, H. (2007) Advanced technologies based on a-Si or LTPS (low temperature poly Si) TFT (thin film transistor) for high performance mobile display. *SID Tech. Digest*, **38**, 1262–1265.
[8] You, B.H., Lee, B.J., Lee, J.H., Koh, J.H., Kim, D.-K., Takahashi, S., Kim, N.D., Berkeley, B.H. and Kim, S.S. (2009) LCD embedded hybrid touch screen panel based on a-Si:H TFT. *SID Tech. Digest*, **40**, 439–442.
[9] Kanda, E., Eguchi, T., Hiyoshi, Y., Chino, T., Tsuchiya, Y., Iwashita, T., Ozawa, T., Miyazawa, T. and Matsumoto, T. (2008) Integrated active matrix capacitive sensors for touch panel LTPS TFT LCDs. *SID Tech. Digest*, **39**, 834–837.

[10] Takahashi, S., Lee, B.J., Koh, J.H., Sito, S., You, B.H., Kim, N.D. and Kim, S.S. (2009) Embedded liquid crystal capacitive touch screen technology for large size LCD applications. *SID Tech. Digest*, **40**, 563–566.

[11] Park, H.-S., Ji, S.-B., Lee, S.-Y., Han, M.-K., Lee, J.-H., Lee, B.-J., You, B.-H. and Kim, N.-D. (2009) A touch sensitive liquid crystal display with embedded capacitance detector arrays. *SID Tech. Digest*, **40**, 574–577.

[12] Destura, G.J.A., Osenga, J.T.M., van der Hoef, S.J. and Pearson, A.D. (2004) Novel touch sensitive in-cell AMLCD. *SID Tech. Digest*, **35**, 22–23.

[13] You, B.H., Lee, B.J., Lee, K.C., Han, S.Y., Koh, J.H., Lee, J.H., Takahashi, S., Berkeley, B.H., Kim, N.D. and Kim, S.S. (2008) 12.1-inch a-Si:H TFT LCD with embedded touch screen panel. *SID Tech. Digest*, **39**, 830–833.

[14] You, B.H., Lee, B.J., Han, S.Y., Takahashi, S., Berkeley, B.H., Kim, N.D. and Kim, S.S. (2009) Touch-screen panel integrated into 12.1-inch a-Si:H TFT LCD. *J. Soc. Inf. Disp.*, **17**, 87.

[15] Hanaoka, K., Nakanishi, Y., Inoue, Y., Tanuma, S. and Koike, Y. (2004) A New MVA-LCD by Polymer Sustained Alignment Technology. *SID Tech. Digest*, **35**, 1200–1203.

[16] Lee, S.H., Kim, S.M. and Wu, S.T. (2009) Emerging vertical alignment liquid crystal technology associated with surface modification using UV curable monomer. *J. Soc. Info. Disp.*, **17**, 551–559.

[17] Lee, Y.-J., Kim, Y.-K., Jo, S.I., Gwag, J.S., Yu, C.-J. and Kim, J.-H. (2009) Surface-controlled patterned vertical alignment with reactive mesogen. *Opt. Express*, **17**, 10298.

[18] Noh, J.-D., Kim, H.Y., Kim, J.M., Koh, J.W., Lee, J.Y., Park, H.S. and Lee, S.H. (2002) Pixel structure of the ultra-FFS TFT-LCD for strong pressure-resistant characteristic. *SID Tech. Digest*, **33**, 224–227.

[19] Kim, H.Y., Seen, S.M., Jeong, Y.H., Kim, G.H., Eom, T.Y., Kim, S.Y., Lim, Y.J. and Lee, S.H. (2005) Pressure-resistant characteristic of fringe-field switching (FFS) mode depending on the distance between pixel electrodes. *SID Tech. Digest*, **36**, 325–327.

Index